RISKS AND CONFLICTS: LOCAL RESPONSES TO NATURAL DISASTERS

Indigenous Fijians Evacuating from the January 2012 Flood in the Ba Watershed, Viti Levu, Fiji. *Source:* Courtesy of Votua Community.

COMMUNITY, ENVIRONMENT AND DISASTER RISK MANAGEMENT VOLUME 14

RISKS AND CONFLICTS: LOCAL RESPONSES TO NATURAL DISASTERS

EDITED BY

ANDREAS NEEF

Centre for Development Studies, University of Auckland, Auckland, New Zealand

RAJIB SHAW

Graduate School of Global Environmental Studies, Kyoto University, Kyoto, Japan

Emerald

United Kingdom – North America – Japan
India – Malaysia – China

Emerald Group Publishing Limited
Howard House, Wagon Lane, Bingley BD16 1WA, UK

First edition 2013

British Library Cataloguing in Publication Data
A catalogue record for this book is available from the British Library

ISBN: 978-1-78190-820-4
ISSN: 2040-7262 (Series)

CONTENTS

LIST OF CONTRIBUTORS

Wojciech Biernacki	Chair of Environmental Sciences, University School of Physical Education in Kraków, Kraków, Poland
Anita Bokwa	Institute of Geography and Spatial Management, Jagiellonian University, Kraków, Poland
Boris Braun	Institute of Geography, University of Cologne, Cologne, Germany
Jarosław Działek	Institute of Geography and Spatial Management, Jagiellonian University, Kraków, Poland
Peter Elstner	Mae Rim, Chiang Mai, Thailand
Konstantin Greger	Graduate School of Life and Environmental Sciences, University of Tsukuba, Tsukuba, Ibaraki, Japan
Umma Habiba	International Environment and Disaster Management Laboratory, Graduate School of Global Environmental Studies, Kyoto University, Kyoto, Japan
Misao Hashimoto	Graduate School of Life and Environmental Sciences, University of Tsukuba, Tsukuba, Ibaraki, Japan
Tomoko Kubo	Faculty of Education, Gifu University, Gifu, Japan
Keisuke Matsui	Graduate School of Life and Environmental Sciences, University of Tsukuba, Tsukuba, Ibaraki, Japan

Shohei Matsuura	International Environment and Disaster Management Laboratory, Graduate School of Global Environmental Studies, Kyoto University, Kyoto, Japan
Michihiro Mashita	Graduate School of Life and Environmental Sciences, University of Tsukuba, Tsukuba, Ibaraki, Japan
Tarun Kumar Mondal	Department of Geography, University of Kalyani, Kalyani, West Bengal, India
Roy Montgomery	Department of Environmental Management, Lincoln University, Lincoln, New Zealand
Farah Mulyasari	International Environment and Disaster Management Laboratory, Graduate School of Global Environmental Studies, Kyoto University, Kyoto, Japan
Andreas Neef	Centre for Development Studies, University of Auckland, Auckland, New Zealand
Annika Salingré	Institute of Geography, University of Cologne, Cologne, Germany
Iven Schad	German Federal Ministry for Economic Cooperation and Development, Bonn, Germany
Rajib Shaw	International Environment and Disaster Management Laboratory, Graduate School of Global Environmental Studies, Kyoto University, Kyoto, Japan
Tom Waldichuk	Geography and Environmental Studies, Faculty of Arts, Thompson Rivers University, Kamloops, BC, Canada
Eberhard Weber	School of Geography, Earth Sciences & Environment, The University of the South Pacific, Suva, Fiji Islands

Toshiki Yamamoto Graduate School of Life and Environmental Sciences, University of Tsukuba, Tsukuba, Ibaraki, Japan

Othniel Yila Stockholm Environment Institute, Africa Centre, Nairobi, Kenya

ABOUT THE EDITORS

Andreas Neef
Centre for Development Studies, University of Auckland, Auckland, New Zealand

Rajib Shaw
Graduate School of Global Environmental Studies, Kyoto University, Kyoto, Japan

Andreas Neef is a professor in development studies at the Centre for Development Studies, University of Auckland, New Zealand. He holds a Ph.D. degree in "Agricultural Sciences (Agricultural Economics and Rural Sociology)" and a postdoctorate degree in "Social Sciences in Agriculture," both from the University of Hohenheim, Stuttgart, Germany. He was previously employed as professor of resource governance and participatory development at the Graduate School of Global Environmental Studies, Kyoto University, Japan. He has extensive field research experience in Southeast Asia, West Africa, the Middle East, and the South Pacific. His research focuses on natural resource governance, land grabbing, participatory approaches to research and development, rural innovation processes, local adaptation to climate change, and post-disaster response and recovery.

Rajib Shaw is an associate professor in the Graduate School of Global Environmental Studies of Kyoto University, Japan. He worked closely with the local communities, NGOs, governments, and international organization, including United Nations, especially in the Asian countries. He is currently the Chair of the United Nations Asia Regional Task Force for Urban Risk Reduction. His research interests are: community based disaster risk management, climate change adaptation, urban risk management, and disaster, and environmental education.

BRIEF INTRODUCTION OF THE SERIES

COMMUNITY, ENVIRONMENT AND DISASTER RISK MANAGEMENT

This series connects academic research to field practice, strengthening the links between the environment, disaster, and community. The series will be developed on field evidences and community practices, and thus will provide specific guides to professionals which are grounded in rigorous academic analysis. The series will have specific focus on community-based disaster risk management, urban environmental management, human security, water community, risk communication, climate change adaptation, climate disaster resilience, and community-based practices.

BRIEF INTRODUCTION OF THE VOLUME

RISKS AND CONFLICTS: LOCAL RESPONSES TO NATURAL DISASTERS

Communities affected by natural disasters are often stigmatized as being passive with regard to disaster prevention, mitigation, and adaptation, waiting only for government assistance in the aftermath of such events. Yet innovative community initiatives and individual livelihood strategies are oftentimes developed in the immediate aftermath of natural disasters, reflecting a high level of energy, creativity, and solidarity. While scrutinizing the potential strength and comparative advantage of such local responses to disasters, this book will analyze and illustrate how individuals, families, and social groups in rural and urban communities perceive natural disasters, their underlying reasons, and their effects on their livelihoods. Key factors determining the degree of vulnerability and resilience of affected communities and social groups will also be elicited. The book will address several key issues related to local disaster responses, such as the relationship between local and external knowledge in addressing disasters, the relevance of institutional, and policy frameworks for community-based disaster response and the role of social capital and collective action in post-disaster recovery.

PREFACE

Natural hazards are believed to cause major disruptions of the functioning of affected communities, often triggering widespread human, material, and economic losses. Yet, at the same time, they often testify to the capacity and energy of local organizations, social groups, and individual citizens to self-organize and to find creative and effective ways to cope in the face of disaster and its aftermath. Following the immediate disaster response and in the course of the long-term recovery process, communities and individual tend to adopt numerous strategies to diversify their resource and livelihood base, to rebuild social networks, and to improve disaster preparedness, often with the assistance of NGOs and other external organizations.

The main questions addressed by the contributions in this volume are as follows:

- How do individuals, families, and social groups in rural and urban communities perceive natural disasters, their underlying reasons, and their effects on their livelihoods?
- What are specific factors that determine the degree of vulnerability and resilience among disaster-affected communities?
- How can external expert knowledge and local wisdom and experience be creatively combined to enhance risk communication and local communities' preparedness for natural disasters?
- What are the most vulnerable groups in local communities and how can their resilience to disasters be strengthened and their long-term livelihoods be secured?
- How can institutional and policy frameworks in support of local disaster risk management be designed?
- What is the role of social capital and collective action in enhancing disaster preparedness, mitigation and adaptive capacity at the local level?
- How can community organizations be linked more effectively with regional, national and international organizations?
- What are the interfaces and possible tradeoffs between individual, community and state responses to natural disasters?

The idea for this book emerged from three sessions on "Local Responses to Natural Disasters" organized by the editors under the theme "Risk and Conflicts" at the 32nd International Geographical Congress held in Cologne, Germany from 26 to 30 August 2012. It comprises five selected papers presented in Cologne and five papers that were added because their content appeared to fit well to the common theme of this book.

The first editor acknowledges the support of the Japan Society for Promotion of Science (JSPS) for a project on "Integrated Assessment of Post-Disaster Recovery of Coastal Communities in Asia-Pacific" (Kakenhi Project Number 24401007) which laid the foundation for this publication project.

The editors hope that this volume stimulates a more informed debate on the various roles that local organizations, community self-help groups and engaged citizens can play in fostering disaster preparedness, response and recovery at the local level if they are provided with the right institutional frameworks and well-focused external assistance.

Andreas Neef
Rajib Shaw
Editors

CHAPTER 1

LOCAL RESPONSES TO NATURAL DISASTERS: ISSUES AND CHALLENGES

Andreas Neef and Rajib Shaw

INTRODUCTION TO THE THEME OF THIS VOLUME

The aftermath of natural disasters poses numerous challenges for communities, organizations, businesses, households, and individual citizens at the local level. Essential infrastructure may be destroyed, basic services are often disrupted, numerous livelihoods are endangered, and the local economy may be in shatters (Berke, Kartez, & Wenger, 1993; Smith, 2001). When natural hazards strike, community organizations have a pivotal role to play in the face of disaster (King, 2007). Prior to the arrival of external assistance, local people have to engage in search and rescue efforts, local organizations need to deploy various types of collective action, and families and individuals have to rely on their social networks to get immediate assistance. Family and community networks tend to be the primary source of disaster relief, whereby "[p]ersonal trust, shared suffering, physical proximity, and long-standing social ties offer a ready-made foundation for cooperation" (Duus, 2012, p. 179). In short, communities, local officials, and ordinary citizens take various measures during and after disasters to reduce disaster risks and ensure quick response and recovery.

Risks and Conflicts: Local Responses to Natural Disasters
Community, Environment and Disaster Risk Management, Volume 14, 1–8

ISSN: 2040-7262/doi:10.1108/S2040-7262(2013)0000014007

Yet, surprisingly, communities affected by natural disasters are often depicted as being passive with regard to disaster prevention, mitigation and responses, and waiting only for government assistance in the aftermath of such events (Blaikie et al., 1994; Morris, 2012). Local people are widely believed to lack the capacity and skills to cope with and recover from disasters. Innovative community initiatives that are developed to address natural disasters are often overlooked or disregarded by external experts, policy makers, and scientists alike. Steward and Donovan (2008, p. 219), on contrast, hold that "[t]he depiction of disaster victims as dazed, confused and helpless individuals, suddenly dependent on external handouts, is a fallacy." A number of attempts have been made recently to give more consideration to local disaster response, preparedness and adaptation and their underlying social and cultural foundations – primarily under such keywords as "community-based disaster preparedness" (e.g., Allen, 2006), "community-based disaster mitigation" (e.g., Shaw, 2006), "social capital" (e.g., Aldrich, 2011; Chamlee-Wright & Storr, 2011), "community organizations" (Yuuko, Pekkanen, & Yutaka, 2012), "social networks" (e.g., Hilfinger, Messias, Barrington, & Lacy, 2012), "social-ecological resilience" (Adger, Hughes, Folke, Carpenter, & Rockstrom, 2005) or "social infrastructure" (Aldrich, 2012).

However, there is still a lack of understanding of how local responses to natural disasters at individual, household and community level are embedded in broader institutional frameworks and wider disaster management efforts that tend to be dominated by outsiders, such as regional government agencies, national decision-makers, and international aid organizations. Another issue that has not been sufficiently scrutinized is how international, national, community, household-level and individual disaster management activities and post-disaster recovery efforts interact in the local "arena." As natural hazards may strike when they are not expected, disaster response and recovery processes are rarely characterized by well-coordinated mechanisms, but often come along with a high degree of "messiness," that is, with coordination problems at various administrative levels and geographical scales and often conflicting views on what is the appropriate response at the local level. Given the diversity of organizations that tend to be involved in post-disaster response and recovery, it seems inevitable that problems of mismanagement and communication failures occur (King, 2007; Steward & Donovan, 2008). Since natural hazards may reach catastrophic levels that exceed the capacity of affected communities to cope by using their own resources, it is crucial to identify ways of more effectively linking organizations and social groups at the local level with regional, national and international assistance. It is obvious that the effectiveness and success of post-disaster response and recovery depends to a great extent on the degree of coordination of external assistance at higher administrative levels with local self-organization. Yet, much scholarly work in disaster management still has a tendency to present the responses to disasters by individuals, households, communities, and external organizations as distinct categories rather than interwoven strands in a complex process (e.g., Paul, 2011).

Another common problem in disaster management research and practice is that decision makers and scientific experts tend to equate "communities" with administrative units at the local level, such as villages, communes, wards, or municipalities, while ignoring the heterogeneity of local stakeholder groups and the various social subunits that may exist at local level and that are characterized by varying degrees of vulnerability, marginalization, and resilience. These subunits – that are often differentiated by class, ethnicity, gender, profession, and age – may also differ widely in their adoption of coping strategies in the wake of disasters (for a comprehensive discussion of what is "community" in the context of natural hazards, see Handmer & Dovers, 2013). Boundaries between these different social subunits and local stakeholder groups may be weakened or strengthened in the direct aftermath of a disaster and in the long-term recovery process that follows.

This volume illustrates and analyses how individuals, families, and social groups in rural and urban communities perceive natural disasters, their underlying reasons, and their effects on their livelihoods. Key factors determining the degree of vulnerability and resilience of affected communities are also elicited. The book addresses several key issues and challenges related to local responses to natural disasters, such as the relationship between local and external knowledge in addressing disasters and communicating risks, the relevance of institutional frameworks and organizational capacity for community-based disaster response, and the pivotal but potentially ambiguous role of social capital and collective action in post-disaster response and recovery. Several chapters illustrate how local resilience can be strengthened by the adoption of various risk diversification strategies, by improved risk communication and by the provision of better information on future hazard risks, for example, through participatory workshops with the most vulnerable groups in the community.

STRUCTURE OF THE BOOK

This volume is divided into two parts: following this introductory chapter, the first part titled "Post-Disaster Response and Recovery at the Local Level" comprises five chapters that deal with issues of direct local responses to disasters, that is, the immediate coping strategies in the aftermath of natural hazards. The second part titled "Adaptation and Risk Reduction Strategies as Local Responses to Disasters" features another set of five chapters that address longer term risk reduction strategies by communities that have been repeatedly affected by disasters and/or that live in extremely risk-prone environments.

Chapter 2, written by Tomoko Kubo and her coauthors, analyses the relationship between community support systems and the behavioral response of residents in Hitachi City, Ibaraki Prefecture, to the Tohoku Pacific Earthquake and Tsunami of March 11, 2011 that claimed the lives of thousands of people in the northeastern part of Japan. By adopting a perceptual–behavioral approach that dismisses positivist quantitative approaches to geographical disaster studies, they aim at providing a deeper understanding of the various responses of particularly vulnerable individuals and communities in the immediate aftermath of a major disaster. The authors argue that the characteristics of local communities were strong, determining factors of the behavior of residents during and after this massive earthquake and tsunami.

In Chapter 3, Roy Montgomery describes and assesses two community-based attempts of local residents in Christchurch, affected by a sequence of damaging earthquakes over a three-year period, to fill the gap that the central government's one-off response and local governments' recovery plans have left in several neighborhoods. The first project "Greening the Rubble" is an initiative of volunteers that sought to turn public and private sites vacated or damaged by the earthquake series into cultural and ecological niches amidst the slowly recovering city. The second project, the "Mt Pleasant Community Response Plan 2012–2013," is a community-based emergency response plan aimed at complementing official civil defence planning. The author scrutinizes these community-based initiatives with regard to risk management, conflict resolution, local participation, and sustainability. He argues that local responses to catastrophic events need to be anticipatory and accommodate a set of factors experienced at the individual and collective level.

Chapter 4, authored by Othniel Yila, Eberhard Weber, and Andreas Neef, demonstrates the significant role of social cohesion and mutual assistance – scientifically expressed in the theoretical concepts of "social capital" and "collective action" – in helping communities in the north-western part of Viti Levu, the largest of the Fijian islands, cope with the devastating consequences of two consecutive flash-floods that hit the downstream part of the Ba River watershed in 2012. The authors discern four strategies that were adopted in the disaster response and recovery process, namely search and rescue, information dissemination, mutual assistance, and socio-commercial cooperation. They also show how disaster recovery efforts contributed to further strengthening social ties within the affected communities (bonding social capital) and helped build linkages with various external aid agencies (bridging social capital) which may prove useful in hedging against future disaster events.

In Chapter 5, Andreas Neef, Peter Elstner, and Iven Schad compare local perceptions and responses to flood disasters in two upland watersheds of North Thailand and Northwest Vietnam. They find that farmers may not be willing to engage in flood mitigation owing to the common perception that the negative impacts of flood events are not primarily a consequence of their own actions, but caused by a bundle of exogenous factors beyond local people's control. The authors argue that local responses to natural hazards are shaped by differential and sometimes conflicting interests and strategies of individual households, social groups at the community level, and state actors. The authors conclude that external actors involved in flood-prevention strategies and policies need to be aware of local people's own causal explanations of flood events and the various trade-offs and conflicts that may exist between collective action toward flood mitigation, individual livelihood strategies, and government agencies' disaster policies and responses.

Chapter 6, written by Umma Habiba and Rajib Shaw, looks into farmers' strategies of coping with drought events, drawing on fieldwork in drought-prone areas of northwestern Bangladesh. After highlighting the causes, types, and impacts of droughts in Bangladesh, the authors elicit various types of adaptive responses to droughts adopted by farmers using their own resources and local knowledge and skills. They find that farmers' coping strategies tend to be reactive rather than proactive and therefore should be complemented by assistance from government agencies, NGOs, and private organizations that needs to be well-targeted to local people's needs and priorities. The authors also call for a greater voice of local communities in formulating drought management plans and implementing drought management projects.

Part II shifts the volume's focus to local responses that are of a more long-term nature and aimed at building adaptive capacity among local communities or particular social or professional groups. In Chapter 7, Tarun Kumar Mondal explores local people's perceptions on natural disasters and their survival strategies by distinguishing between the major occupational groups in the Sunderban Delta region of West Bengal, India. He identifies a considerable variation in the perception of disaster risks and disaster impacts between the various occupational groups, which is also reflected in their respective coping strategies. Drawing on these empirical results, he argues convincingly that such occupation-specific disaster risk perceptions and survival strategies need to be built into disaster management planning and disaster-related policy making.

In Chapter 8 Jarosław Działek, Wojciech Biernacki and Anita Bokwa analyze the linkages between the degree of social capital and local response to floods among communities in Southern Poland by taking a historical – geographical approach. While they identify a relatively high level of passivity toward flood prevention in general, they also find evidence of spatial differences which they explain by a combination of historical, cultural, and social factors, such as community size, historical trajectories and different regional values and norms toward collective action. They conclude that the assessment of local flood risks needs to build on an improved understanding of how to harness local network capacities toward strengthening community resilience and informing local flood risk governance strategies.

Chapter 9, authored by Rajib Shaw and Shohei Matsuura, emphasizes the important role of schools in enhancing disaster preparedness at the local level and in reducing the adverse social impacts of disasters. Drawing on the experience from various Japanese cities, the authors argue that schools can serve as multifunctional community hubs as they do not just provide education for children, but also may function as facilities that strengthen social cohesion in a community and revitalize sociocultural identity. They propose a set of policy actions to enhance pre-disaster preparedness and ensure disaster risk reduction through improved school-community linkages.

In Chapter 10, Farah Mulyasari and Rajib Shaw elicit the potential role of Community-Based Society Organizations (CBSOs) in enhancing risk communication in disaster-prone areas of Indonesia by a comparative analysis of three types of groups (women's group, youth union, and faith-based organizations). They identify various risk communication processes, such as dissemination of disaster risk information and conveyance of early warnings through dense social networks. They conclude that CBSOs have the potential to act as mediators between government and community through bridging communication gaps, if they can overcome several challenges, such as providing trustworthy information, and clearly define responsibilities and build competencies along the information channels.

Chapter 11, written by Annika Salingré and Boris Braun, discusses the value of grassroots workshops to communicate climate change adaptation and disaster risk reduction strategies among the urban poor in hazardprone slums of Dhaka, Bangladesh. The authors argue that formerly successful coping and adaptation strategies among slum dwellers may be challenged by new disaster threats induced by global warming. They describe experience gathered and lessons learned from a series of participatory workshops with adults and children in the context of a German–Bengali action research project and suggest a number of strategies to overcome challenges of participation fatigue, psychological distance and measurement of success.

The aim of this volume is to go beyond simplistic notions of "community-based" disaster management and to provide sound empirical evidence for the diversity of strategies, conflicts, and interests at the local level that may determine the success or failure of disaster risk reduction and post-disaster response and recovery.

REFERENCES

Adger, W. N., Hughes, T. P., Folke, C., Carpenter, S. R., & Rockstrom, J. (2005). Social-ecological resilience to coastal disasters. *Science*, *309*, 1036–1039.

Aldrich, D. P. (2011). The externalities of strong social capital: Post-tsunami recovery in Southeast India. *Journal of Civil Society*, *7*(1), 81–99.

Aldrich, D. P. (2012). Social, not physical, infrastructure: The critical role of civil society after the 1923 Tokyo earthquake. *Disasters*, *36*(3), 398–419.

Allen, K. (2006). Community-based disaster preparedness and climate adaptation: Local capacity-building in the Philippines. *Disasters*, *30*(1), 81–101.

Berke, P. R., Kartez, J., & Wenger, D. (1993). Recovery after disaster: Achieving sustainable development, mitigation and equity. *Disasters*, *17*(2), 93–109.

Blaikie, P., Cannon, T., Davies, I., & Wisner, B. (1994). *At risk: Natural hazards, people's vulnerability and disasters* (1st ed.). New York, NY: Routledge.

Chamlee-Wright, E., & Storr, V. H. (2011). Social capital as collective narratives and post-disaster community recovery. *The Sociological Review*, *56*(2), 266–282.

Duus, P. (2012). Dealing with disaster. In J. Kingston (Ed.), *Natural disaster and nuclear crisis in Japan: Response and recovery after Japan's 3/11* (pp. 175–187). New York, NY: Routledge.

Handmer, J., & Dovers, S. (2013). *Handbook of disaster policies and institutions: Improving emergency management and climate change adaptation* (2nd ed.). New York, NY: Routledge.

Hilfinger Messias, D. K., Barrington, C., & Lacy, E. (2012). Latino social network dynamics and the Hurricane Katrina disaster. *Disasters, 36*(1), 101–121.

King, D. (2007). Organisations in disasters. *Natural Hazards, 40*, 657–665.

Morris, J. F. (2012). Recovery in Tohoku. In J. Kingston (Ed.), *Natural disaster and nuclear crisis in Japan: Response and recovery after Japan's 3/11* (pp. 33–50). New York, NY: Routledge.

Paul, B. K. (2011). Environmental hazards and disasters: Contexts, perspectives and management. Oxford: Wiley-Blackwell.

Shaw, R. (2006). Critical issues of community based flood mitigation: Examples from Bangladesh and Vietnam. *Science & Culture, 72*(1), 62–72.

Smith, K. (2001). *Environmental Hazards: Assessing Risk and Reducing Disaster* (3rd ed.). New York, NY: Routledge.

Steward, I., & Donovan, K. (2008). Natural hazards. In S. Buckingham & M. Turner (Eds.), *Understanding environmental issues* (pp. 207–234). London: Sage.

Yuuko, K., Pekkanen, R., & Yutaka, T. (2012). Civil society and the triple disaster. In J. Kingston (Ed.), *Natural disaster and nuclear crisis in Japan: Response and recovery after Japan's 3/11* (pp. 78–93). New York, NY: Routledge.

PART I
POST-DISASTER RESPONSE AND RECOVERY AT THE LOCAL LEVEL

CHAPTER 2

THE RELATIONSHIP BETWEEN COMMUNITY SUPPORT AND RESIDENT BEHAVIOR AFTER THE TOHOKU PACIFIC EARTHQUAKE: THE CASE OF HITACHI CITY IN IBARAKI PREFECTURE

Tomoko Kubo, Toshiki Yamamoto, Michihiro Mashita, Misao Hashimoto, Konstantin Greger, Tom Waldichuk and Keisuke Matsui

ABSTRACT

Drawing on a case study in Hitachi City, Ibaraki prefecture, this chapter aims to analyze the relationship between community support and the behavior of residents after the Tohoku Pacific Earthquake in the regions affected by the disaster. The chapter will examine residents' behavior and the community's roles by way of the following process: (1) We will

Risks and Conflicts: Local Responses to Natural Disasters
Community, Environment and Disaster Risk Management, Volume 14, 11–42

ISSN: 2040-7262/doi:10.1108/S2040-7262(2013)0000014008

review Japan's natural disaster prevention regimes; (2) we will examine the result of a field survey conducted in Hitachi City detailing the city's natural disaster prevention procedures and the operation of some neighborhood evacuation sites; (3) the behavior of residents following the earthquake is analyzed. In this part, questionnaires were sent to 2000 households, of which 492 (24.6%) were collected and used for this analysis. The earthquake and tsunami destroyed lifelines such as water supply for several days in the city. According to the city, a total of 65 buildings were judged to be in dangerous condition, 251 as requiring care, and 478 were only partially damaged. The most serious damage was found mainly in the city's coastal areas, where a total of 85 houses were entirely or partly damaged, and 483 houses were flooded above the floorboards by the tsunami. On March 11, a total of 69 evacuation sites opened, and 13,607 residents rushed into them. After the disaster, residents initially tried to go back to their homes. Depending on the damage done, they either stayed there or moved to a relative's or friend's house, or to a neighborhood evacuation site. Due to the failure of the lifelines, transportation systems, and the damage caused by the disaster, most residents had to stay within an area more limited than usual, around which they could walk or ride by bicycle. Residents had only the human and physical resources of their neighborhoods. Therefore, the characteristics of their local communities affected how residents behaved during and after the earthquake.

Keywords: Local response; behavior; evacuation sites; the Tohoku Pacific Earthquake; Hitachi City

INTRODUCTION

Background of the Study

The Tohoku Pacific Earthquake and Tsunami of 2011 caused incalculable damage in Eastern Japan. Both the Japanese government and scholars in many academic fields have been working for a rapid recovery from this disaster. Big earthquakes affect local economies and communities, thus disrupting the daily lives of residents. Urban reconstruction can, however, revive those damaged, and many actors, including the national government, local authorities, non-profit organizations (NPOs), and local residents, have been evaluating restoration plans (Edgington, 2010).

In Japan, local communities play particularly important roles during and after disasters. Local governments designate and operate evacuation sites, usually elementary and junior high schools, and residents are trained in evacuation drills to move to them. Local communities and school personnel actively engage in their practical functioning and also assist in the reconstruction of neighborhoods (Ishii, Yamazaki, Namai, Uchida, & Okazawa, 1996). During and after disasters, most residents have to stay in smaller areas than usual and travel on bicycles rather than in automobiles, since the physical distribution of products, including gasoline, has completely stopped and roads are damaged. At the same time, public services and transportation are severed. Thus, the activity fields of residents are confined to walking distances, a limitation that stimulates the formation of communal groups (Fujioka, 1996). Heeren (1999) mentions that disaster victims build bonds in their neighborhoods because they simultaneously share the same strong emotions. A disaster can be an opportunity for residents to discover their neighborhoods. Despite this seminal fact, the relationship between community support systems and the behaviors of residents after an earthquake has not been sufficiently examined.

Studies of the 1995 Great Hanshin-Awaji Earthquake in the Kansai region of Japan have shown that the degree of damage was influenced by the area's residential structure (Kagawa, 1995; also see Table A.1 in the appendix). Ishii et al. (1996) demonstrate that earthquake victims tend to be the weak, such as elderly females, and that an area's residential characteristics, such as the types of housing and building materials, determine the levels of damage (Fukutome, 1999). In the Japanese housing system, those with private means dwell in disaster-resistant houses on hillside, while those without them tend to live in social housing in disaster-prone areas (Hirayama, 2000). Housing structures are closely related to social structures and community characteristics; thus, opportunities and damages are unevenly distributed in disaster-affected regions.

To examine the successful urban reconstruction that took place after the Tohoku Pacific Earthquake and Tsunami, we address the following questions in this chapter: (1) What was the actual damage caused by the earthquake? (2) How did the community proceed after the earthquake? (3) Did community support systems affect the behavior of residents, and if they did, how did they affect it?

The problem of sketchy information was pointed out after the Great Hanshin-Awaji Earthquake. Although the disaster spread throughout the

Kansai region, information about damage was limited to the core areas (Kanasaka, 1995). However, detailed facts and field surveys of the surrounding regions were required for reconstruction planning and damage containment, since destruction of electricity, water services, and core transportation systems was widespread, indicating that more complete disaster damage prevention plans were necessary (Kagawa, 1995).

Methodology of the Study

This study analyzes the relationship between community support systems and the behavioral responses of residents after the Tohoku Pacific Earthquake in the zones surrounding the disaster. Our study area is Hitachi City in the Ibaraki prefecture, where the earthquake and tsunami affected the daily lives of residents.

To achieve our objective, we adopted a perceptual and behavioral approach to geography. It focuses on human perceptions of the environment and the decision-making processes that affect behavioral responses.[1] This method opposes the "quantification revolution" in academic research. Its main interests are human behavior, including the perception of natural disasters, consumption patterns, residential choices, the evaluation of residential environments, and the image perceptions of places (Gold, 1980; Okamoto, 1998). The perceptional-behavioral approach contributes to the understanding of human actions in natural disasters (Ganzawa, Kito, & Sadakata, 1995; Okamoto, 1998; Wakabayashi, 2009).

In addition, perceptional-behavioral approaches permit a deep understanding of disadvantaged social groups, such as elderly or handicapped persons, children, women with infants, and foreigners (Okamoto, 1998). As many studies have mentioned, such groups are more vulnerable to natural disasters (Bolin & Stanford, 1998; Ishii et al., 1996; Wisner, 1998), but their needs are rarely reflected in services or disaster-mitigation policies (Takeda, Tamura, & Tatsuiki, 2003). We deal with the behavioral responses of residents after a disaster, placing an emphasis on community support systems that strongly influence their reactions. Resident perceptions of their neighborhoods or residential environments are also discussed.

This chapter examines resident behavioral responses and community roles in the following process. First, we review Japan's natural disaster prevention regime, focusing on school education (including evacuation drills), community preparation for emergencies, and local government disaster responses. Second, we examine the results of a field survey conducted in

Hitachi City that details that community's natural disaster prevention procedures and the operation of some neighborhood evacuation sites. Third, the behavioral responses of residents following the earthquake are analyzed through a mail survey and personal interviews. Questionnaires were sent to 2000 households, of which 492 (24.6%) responded. The questionnaires collected information on (1) the extent of damage to residences, (2) the types of housing (including building materials and construction periods), and (3) resident behavioral responses after the earthquake.

JAPAN'S NATURAL DISASTER PREVENTION REGIME

The National and Local Governmental Regimes for Disaster Prevention

Japan is located on the Circum-Pacific Seismic Belt, which is more popularly called the "Ring of Fire"; geographical, topographical, and meteorological conditions make it one of the most disaster-prone countries in the world. In addition, the nation's population and economic activities are concentrated in highly urbanized areas, which are intensely and intricately exploited and where there are very few open spaces (Mather, Karan, & Iijima, 1998). Japanese disaster management systems have been developed and modified after several large-scale natural catastrophes. In the 1940s, several natural disasters such as Typhoon Makurazaki (1945), the Nankai Earthquake (1946), Typhoon Catherine (1947), and the Fukui Earthquake (1948) caused serious damage (see Table A.1 in the appendix). To permit the recovery from disasters and to provide aid to victims, the Japanese government enacted the Disaster Relief Act (in 1947, after the Nankai Earthquake), the Flood Control Act (in 1949, after the two great typhoons), and the Building Standard Law (in 1950, after the Fukui Earthquake). The major Japanese disaster management acts, plans, and systems were established following Typhoon Ise-wan (1959), the Great Hanshin-Awaji Earthquake (1995), and the Niigata-Chuetsu Earthquake (2004) (Cabinet Office, Government of Japan, 2011; also see Table A.1 in the appendix).

In the Japanese reconstruction management system, the national government has taken responsibility for the reclamation of urban infrastructures and transportation systems and the removal of debris. This charge was established in the period of rapid economic growth, when national or local governments undertook much urban infrastructure construction and development in order to attain the levels of Western cities. Other objectives such as the recovery of damaged communities or the revitalization of local economies were assumed by individuals, NPOs, or private enterprises (Edgington, 2010). Generally, the Japanese people have expected little governmental support, even during or after great disasters (Ozerdem & Jacoby, 2006). A lack of communication exists between national or local governments and citizens in Japan; in fact, it is rare for government officials to take the views of local people into account (Pekkanen, 2006; Schwartz & Pharr, 2003). For disaster-prevention, Japanese planners have developed disaster-proof construction techniques, such as base-isolating buildings or embankments. In contrast, insurance systems for disasters or support systems for victims have been weak in Japan (Murakami, 1996).

Japanese reconstruction management has been characterized by top-down or hierarchical systems (Edgington, 2010). For example, the early warning system for emergencies starts with national institutions (Japan Meteorological Agency, the national government ministries, and the agencies related to disaster management), which pass orders to local governments. These local governments then convey disaster information to their citizens (Cabinet Office, Government of Japan, 2011). When disasters occur, information or orders thus move along this top-down system. For example, during the Great Hanshin-Awaji Earthquake in 1995, local governments officially took part in the management of evacuation sites (obtaining information or support from the national government and distributing it to local sites), but substantial parts of the recovery operation were carried out by school personnel or local communities (Ishii et al., 1996).

Although reconstruction processes are led by the Japanese government or local authorities, the voices of residents became increasingly important after the 1995 Kobe earthquake, which utilized *machizukuri* methods, that is, those that relied on local residents and specialists, such as architects, to create neighborhood rebuilding plans, in addition to traditional top-down methods, such as land readjustment and urban reconstruction (Edgington, 2010).

Evacuation Drills and Community Preparation for Natural Disasters in Japan

Under the Disaster Countermeasures Basic Act, which was adopted in 1961 after Typhoon Ise-Wan, obligatory disaster reduction drills were established in Japan. After the Great Kanto Earthquake of 1923, the September 1 was designated as "Disaster Reduction Day"; on it each year, drills to mitigate natural catastrophes are conducted by governmental institutions, schools, other educational facilities, private enterprises, and local communities (Cabinet Office, Government of Japan, 2011). Through these drills, the Japanese are trained to engage in proper actions in emergencies. Edgington (2010, p. 28) holds, however, that "these exercises may be viewed as symbolic – as 'collective exorcism' – rather than as genuine efforts to train communities in disaster preparation." Indeed, it is unclear if such drills involve sincere participation, a point that will be taken up later in this chapter.

According to Kunugiza and Matsui (2005), when a tsunami hits a residential area just after an earthquake, as a tsunami reached three minutes after the earthquake to the southwest off Hokkaido Earthquake in 1993, top-down information systems cannot work efficiently; it takes at least four minutes until a tsunami warning is broadcast by television or radio. These scholars, therefore, insist that disaster reduction education must be designed to alter human perceptions of risk and to aid in correct decision making. Most university students or residents, however, did not make the "right" choices in emergencies because their discernments were affected by media messages indicating that "helping people die should be praised more than quick survival evacuations" (Kunugiza & Matsui, 2005). In all, 201 people died and 28 disappeared in the 1993 earthquake and tsunami southwest off Hokkaido (Ganzawa et al., 1995). With this tragic background in mind, it is important to realize that local communities, school personnel, and individuals are capable of making informed survival decisions. Evacuation drills in communities, schools, and offices are expected to better emergency decision making.

EARTHQUAKE DAMAGE AND DISASTER PREVENTION IN HITACHI CITY

Geography of Hitachi City

In this section, we examine the disaster-prevention plans of Hitachi City and its reaction to the 2011 earthquake and tsunami. Hitachi City, which is located in northern part of Ibaraki prefecture, has a population of about 200,000. It developed as a mining and manufacturing town at the turn of the last century, when a large part of its land area was in the past covered by copper and pyrites mines. Later, it was widely known as the company town of the Hitachi Ltd, an engineering and electric company. Although the Hitachi Ltd moved its head office to Tokyo, it maintains branch offices and many factories in Hitachi City, about 40% of whose residents work for the company; others work in the fishing, tourism, and agricultural industries, or in manufacturing.

The topography of Hitachi City (Fig. 1) reveals its disaster-prone characteristics. Most of its surface is mountainous; its people live on a narrow, rather low coastal plain. The distance between the coastline and mountainous areas is usually less than five kilometers. In a tsunami occurs, a large part of the residential areas is easily inundated, and limited evacuation roads can cause panic. The topography causes chronic traffic congestion; there is one Joban highway route and three national roads, the south to north Route 6 and Route 245, and the west to east Route 293. In addition, the JR Joban line runs south to north and connects the city to Tokyo and the Tohoku region. Most large housing estates (more than 100 housing units) were built on hillsides from the 1960s onwards as residences for mining workers and Hitachi staff (Fig. 2).

Hitachi's Condition After the Disaster

According to the classic model of Kates and Pijawka (1977), the reconstruction processes following a natural disaster is composed of four phases: the emergency, restoration, replacement/reconstruction, and developmental/reconstruction periods. Hitachi City, which was damaged by the recent earthquake and tsunami, is now in the final period of this model. Although coastal areas were seriously devastated by the tsunami, most of the hotels located in them reopened at the end of September 2011, just six months after the calamity. Fishing and related industries in Hitachi City have not, however, completely recovered, since the levels of radioactive substances in marine products remained higher than usual in 2012. The fishery association of the Ibaraki prefecture has established standards for these substances that are stricter than the national ones, hoping to avoid the harms of rumor that its products contain dangerously elevation concentrations of radioactivity (*fuhyo-higai*). Gradually, fishery industry starts their business as the radioactivity levels decline.[2] Thus, the people and industries of Hitachi City have suffered from both the direct and immediate damage of the earthquake and tsunami and from the long-term effects of the nuclear accident that was triggered by them.

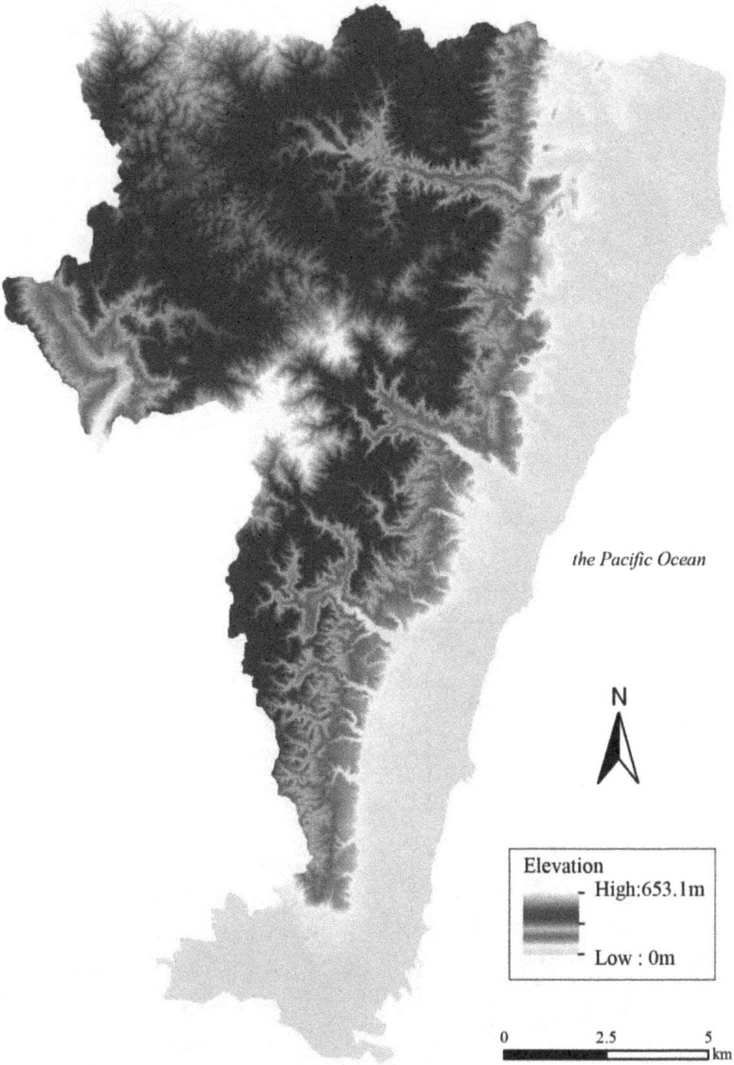

Fig. 1. Topography of Hitachi City.

Local communities, urban structures, and socio-economic or cultural conditions are regarded as important factors in determining the recovery and reconstruction processes (Alexander, 1993; Massard-Guilbaud, 2002; Mileti, 1999) following natural calamities. In the following sections, we

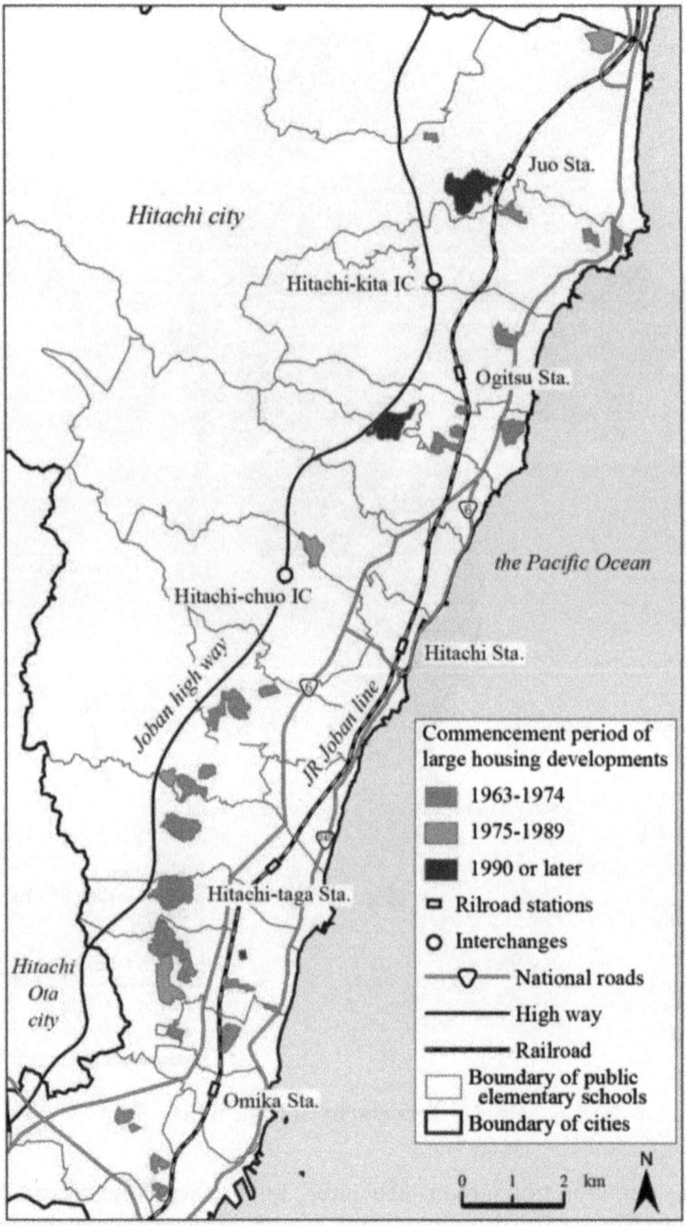

Fig. 2. Distribution of Large Housing Estates in Hitachi City (1963–2006). *Note*: Adapted from Hashimoto et al. (2012). *Source:* Hitachi City Urban Planning Materials (2006).

treat Hitachi as a case study, focusing on the resilience of the city; we explore, in particular, its prevention plan for natural disaster damage, the material and human harms caused by the earthquake and tsunami of 2011, and community responses to them.

Prevention of Disaster Damage

Hitachi City has paid great attention to the preclusion of nuclear emergencies since the Tokai-mura JCO Nuclear Accident in 1999, which follows the Fukushima Daiichi nuclear disaster of 2011 in gravity.[3] Adjacent to Tokai-mura in the south, some of its residents, who are employed in these nuclear facilities, were affected by the JCO incident; as a result, the residents of Hitachi City have greater knowledge of nuclear facilities than the inhabitants of other Japanese areas.

Hitachi City's Act on the Prevention of Emergencies provides for four nuclear emergency advisors in a disaster-countermeasures office. The municipality has been distributing disaster radios to residents since 2000, although 26,000 households have not yet received them (May 2011). It also built broadcasting towers to communicate important information to residents during nuclear emergencies. Hazard maps for natural disasters, including tsunamis, have been published; evacuation sites are designated on them, and evacuation drills are conducted at least once a year by educational institutions, local communities, and companies.

Damage in Hitachi City

The earthquake destroyed major public infrastructures and services, such as the water supply, sewage, electricity, and gas for several days in the city; the water supply was not restored until the end of March in some areas. In a large part of the municipality, houses were damaged only slightly, and roofs and walls only partially. According to the city's report from May 22, a total of 65 buildings were assessed as dangerous, 251 judged as requiring care, and 478 listed as partially damaged. The most serious destruction occurred in the city's coastal areas, where a total of 85 houses were entirely or partly damaged and 483 others were flooded above their floorboards by the tsunami. The Japanese Reconstruction Agency reported that there were 13 people dead by the Tohoku Pacific Earthquake and Tsunami in Hitachi City (March 31, 2013).[4]

On March 11, a total of 69 evacuation sites opened in which 13,607 residents found shelter (Figs. 3 and 4). In addition to the designated evacuation sites, high schools and local community centers were hastily used as shelters. Main roads and the JR Joban line were rendered impassible in several places by the earthquake and subsequent tsunami; people were stranded and rushed into the evacuation sites in the central area of the city, two of which officially accepted evacuees from the Fukushima prefecture (one remained open until March 26 and other until April 11). Local community residents assisted officials at each site, as detailed in the next section.

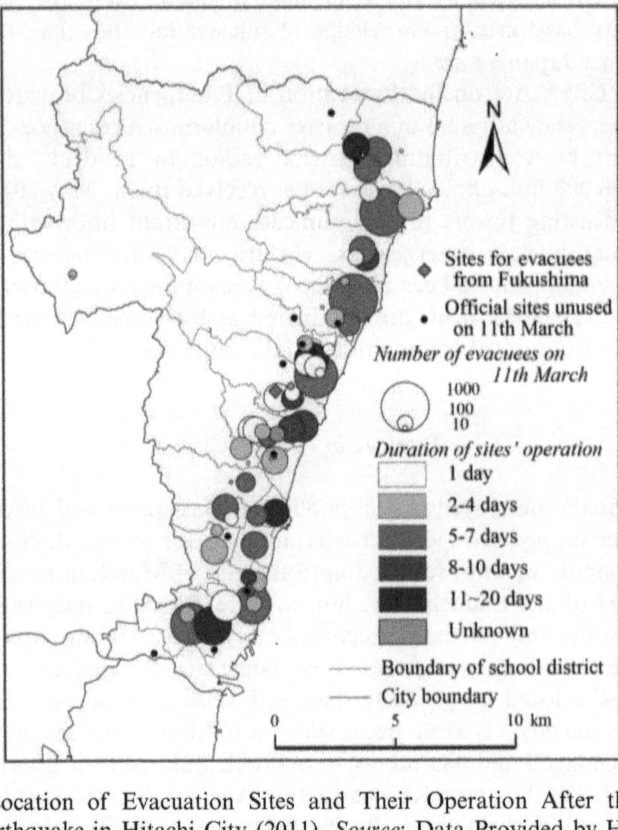

Fig. 3. Location of Evacuation Sites and Their Operation After the Tohoku Pacific Earthquake in Hitachi City (2011). *Source*: Data Provided by Hitachi City Administration.

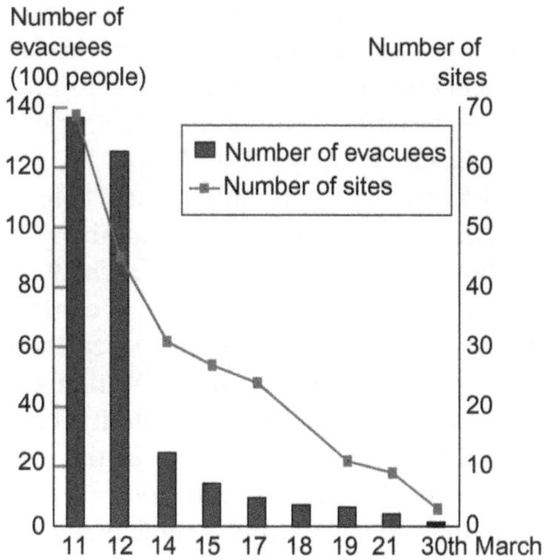

Fig. 4. The Number of Evacuation Sites and Their Evacuees After the Tohoku Pacific Earthquake (March 11–30, 2011) in Hitachi City. *Source*: Data Provided by Hitachi City Administration.

RESPONSE OF LOCAL COMMUNITIES: OPERATION OF EVACUATION SITES

An Old Housing Estate

First, we will describe the case of a housing estate on a hillside area of the city. Developed in 1973, it consists mainly of detached owner-occupied houses. Most of the residents, initially married couples with children, moved into the area when sales began. Over time, children left parental homes and the estate's population fell by one-third; 22.5% of its inhabitants are elderly (January 2010). In recent decades, rental apartments, preferred by young families, and public housing went up in this area. This newer construction has mitigated the rapid aging of the overall community.[5] Since most of the residents shared similar household characteristics and lived in the estate for more than 30 years, community activities were quite established and functional. From 1984 to 1989, the estate was selected to be a model community by the Ministry of Internal Affairs and Communication for community promotions. Later, it won official commendations.

Housing damage was not serious in this area, but the earthquake impaired its public and private services. Residents who lacked food, water, heating, or non-electrical cooking facilities moved into the area's evacuation sites. In addition to the designated site (a gym in an elementary school), a community center was hastily employed. On March 11, 390 residents moved into the school and 105 into the community center (Fig. 5). On March 15, when the area's electricity was restored, the numbers fell to 86 and 30, respectively. Finally, both sites were closed on March 19.

Table 1 indicates the assistance offered by local communities in evacuation site operation. About 25 minutes after the earthquake, community members started patrolling their neighborhood. Within 45 minutes, they set up a local disaster headquarter in the community center. It was soon moved to the city's designated evacuation site. In the evacuation sites, the local community provided (1) food to both evacuees and residents who could not leave, including the elderly and handicapped; (2) medical examinations and exercise instructions, assisted by the area's volunteer medical staff; (3) drinking and waste water; and (4) information on the earthquake.

Thanks to an evacuation drill in November 2010, members of the local community easily assumed their roles during the emergency. Along with the employees of local companies and staffs of medical facilities, they had been participating in annual evacuation drills for years. The community's participation resulted in a well-organized and successful operation.

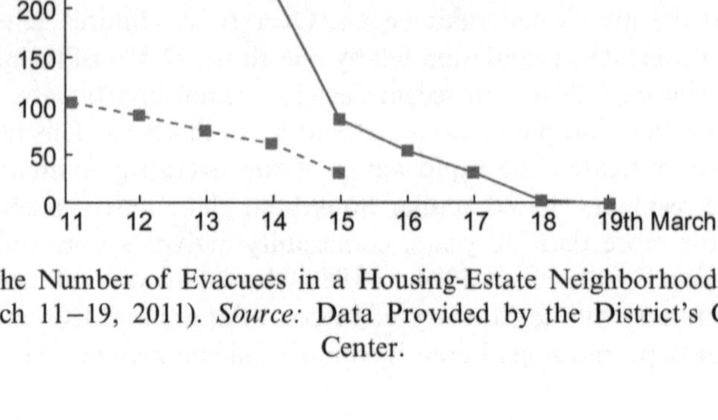

Fig. 5. The Number of Evacuees in a Housing-Estate Neighborhood in Hitachi City (March 11–19, 2011). *Source:* Data Provided by the District's Community Center.

Table 1. Community Support After the Tohoku Pacific Earthquake in a Housing Estate of Hitachi City (March 2011).

Date	Time	Activities by Community Members and Other Activities
11	14:46	Occurrence of the earthquake
	15:10	A patrol by community members (taking care of injured residents, etc.)
	15:30	Set up "local disaster headquarter" in the district's community center
		Meeting of community members; collecting and distributing disaster information
		Visiting residents who ask for support (193 persons); moving them to the evacuation site
		(Community members in charge of welfare ⇒ Welfare officers)
	16:00	Establishment of evacuation sites (community members and city staff)
	16:55	105 evacuees for elementary school, 105 for small community center
15	18:50	Closure of the evacuation site in local community center
		A large part of the district's electricity restored
18	14:00	Main evacuation site was moved to the district's community center
19	16:00	Closure of the evacuation site in the district's community center
21	9:35	Water supply restored in most houses of the district

Source: Data provided by the district's community members.

Electricity service was restored on March 15, and gas, water supply, and sewage services restarted within 10 days.

A Coastal Settlement

The case described here is that of a coastal settlement affected by the tsunami (Fig. 6). At the time of the disaster, most of its residents had lived there for generations and had close connections with each other. They tended to depend on their private networks rather than on their local community. Many worked as fishermen or ran coastal inns. However, hillside areas in the same neighborhood show completely different features; most of their inhabitants, who worked for Hitachi Ltd or its related businesses, had been in residence for less than 30 years. This diverse resident mix resulted into unsuccessful operations at the site.

On Hitachi City's tsunami-hazard map, this coastal settlement and its hillside neighborhoods shared the same designated evacuation sites which were built on a hill. The post-earthquake condition of these sites was, however, inadequate; some of their windows had been broken, allowing cold

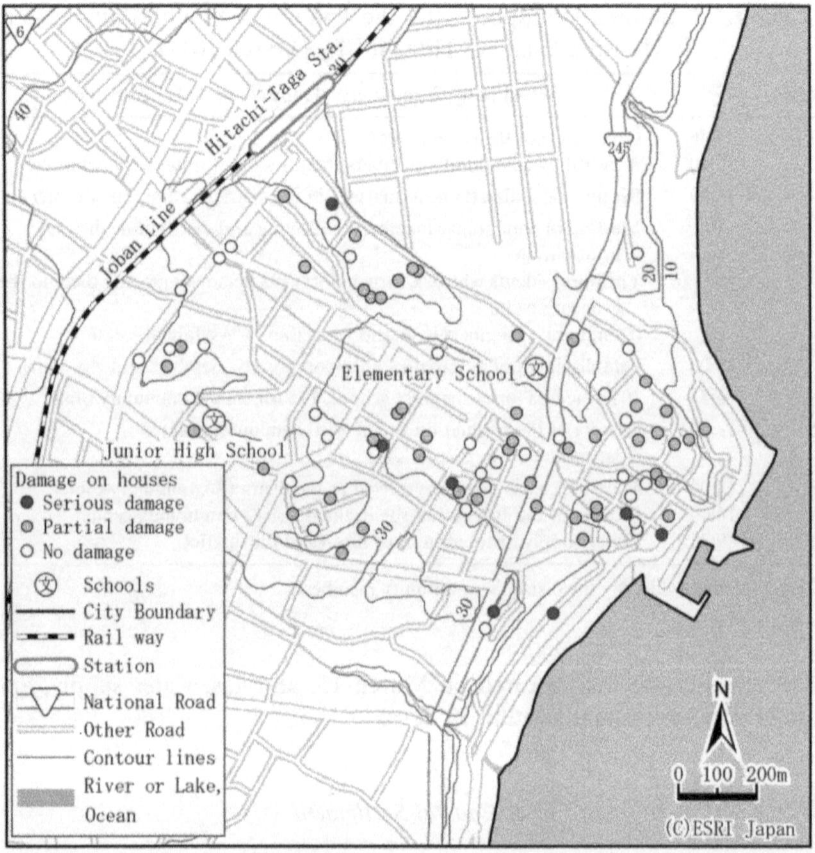

Fig. 6. Damages on Houses After the Tohoku Pacific Earthquake in a Coastal Settlement of Hitachi City (2011). *Source*: Authors' Questionnaire Survey, and Fundamental Geospatial Data.

wind to enter, and they lacked blankets and heat. Although the hillside community tried to support the sites' operation, they found the provisioning of food and water arduous (Table 2). On March 11, an inn with a stockpile of food offered rice to the local community, and volunteers from the hillside community helped distribute them to evacuees. Although members of the local community assisted in these operations, their roles were minimal. In this neighborhood, electricity supply was restored on March 16, and gas, water supply, and sewage began on March 22. The renewal of drinking water was the most difficult task for residents.

Table 2. Community Support After the Tohoku Pacific Earthquake in a Coastal Settlement of Hitachi City (March 2011).

Date	Activities by Community Members and Other Activities
11–20	Volunteer activity by community members and residents
11	Distribution of food (materials from nearby inn and residents)
12	Distribution of food (volunteers from residents) and water
13	Distribution of water for toilets; use of power generators for lighting; visits to elderly people by the district's welfare officer
18	Volunteer work for the victims of tsunami in the district

Source: Data provided by the district's community members.

A large number of coastal residents who suffered serious harm from the tsunami depended on their relatives or friends rather than on the local community. In the coastal settlement, where people had strong economic and family ties, residents preferred staying at home and helping each other rather than moving to evacuation sites. Although they had earlier conducted disaster reduction drills for fires (in which all community members participated) and tsunamis (which were restricted to coastal residents), the evacuation sites did not succeed.

As Alexander (1993) points out, original area conditions or problems can be magnified after disasters. In Hitachi, if a local community was well organized and had its own disaster-response systems before the calamity, it successfully organized evacuation sites and managed the distribution of food, water, and information to residents. In comparison, a community divided by conflicts, some invisible, was far less effective in its supervision of evacuation sites and its reconstruction process. We confirmed the imperceptible split between fishery residents, who preferred traditional, familial, and friendship support bonds, and white-collar new residents, who joined community activities.

BEHAVIORAL RESPONSES OF RESIDENTS AFTER THE EARTHQUAKE

Patterns of Resident Behavioral Responses

The patterns of the behavioral responses of household heads were classified into six categories, which appear in Table 3.

Table 3. Patterns of Household Head Responses After the Tohoku Pacific Earthquake in Hitachi City (2011).

Types of Behavioral Responses	Number	Percentage
(1) Stayed at home	335	68.1
(2) Moved to evacuation sites	4	0.8
(3) Stayed at friends' or relatives' houses	24	4.9
(4) Went to work	26	5.3
(5) Were stranded	66	13.4
(6) A mixture of several types, and others	16	3.3
(7) Unknown	21	4.3
Total	492	100

Source: Authors' questionnaire survey.

- Stayed at home: This category refers to those that went home on March 11 and stayed a long time after the earthquake. A total of 335 household heads (68.1%) followed this pattern. They tended to work either in Hitachi City or in neighboring towns and did not use public transportation to travel to their offices. They remained at home for about one week or longer, exiting only to find well water and open grocery stores or to meet friends or relatives.
- Moved to evacuation sites: Those that moved to evacuation sites after the earthquake fell into this category. Only four household heads (0.8%) moved to evacuation sites. We should mention that spouses, usually females, were more likely to use evacuation sites after the earthquake.
- Stayed at friends' or relatives' houses: In all, 24 respondents (4.9%) stayed at friends' or relatives' houses because their own homes were damaged by the earthquake or tsunami.
- Went to work: This category concerns 26 respondents (5.3%) who went to work soon after the earthquake. Most of them lived in undamaged houses in the new housing estate built in a hillside area. Using the restored public transportation systems or their own cars or bikes, they quickly resumed their work.
- Were stranded: This group includes 66 household heads (13.4%) who were stranded for several days after the failure of transportation systems. They tended to work in Tokyo or were on visits; thus, they could not return home on March 11.
- Mixture of several types: This category is a mixture of the other five types. It comprises 16 answers (3.3%). A total of 21 households (4.3%) did not answer this question.

Effects of Community Support on the Behavior of Residents

In addition to the damage caused by the earthquake and tsunami, the characteristics of the local communities determined evacuation sites operations, thus affecting resident's behavioral responses. This observation is demonstrated in the following three cases.

A Case of a Young Family in a Housing Estate

In the old housing estate, the local community took a leading and successful role in operating the evacuation sites. This helped young residents who had never participated in such activities understand the importance of local community participation. Some young residents moved to the evacuation sites after the earthquake and had opportunities to join in volunteering. In the authors' interviews, these residents mentioned that they had gradually developed an attachment to the local community; this development could help in the building of a sustainable communal spirit. A model of a young family's behavior in this neighborhood in the aftermath of the disaster is depicted in Fig. 7.

The family members are the household head, his spouse, and their little girl; they had never joined in community activities before the earthquake. Just after the earthquake, the wife and her child moved to the evacuation site, and they stayed there until March 23. As their little girl trembled with fear whenever small shakes caused creaks in its house, the family had later moved into the house of the child's grandparents.

In the evacuation site, the wife joined in volunteer work, such as preparing food for evacuees; community members led healthy evacuees in such activities in order to supplement the food or water supplies distributed by the local government. In addition, young mothers with small children helped each other so that they could feel comfortable in the evacuation site. These experiences caused a major change to the wife's mind, who now says, "I want to raise my child in this neighborhood. She will go to an elementary school here. Now, I wish to join community activities much more than before. The local community contributes to our healthy and comfortable daily lives."

A Case of a Family of a Coastal Settlement

In the coastal settlement, by contrast, residents tended to depend on their own personal networks. Many residents settled there generations ago, and their relatives and friends lived nearby. Although they could have used the

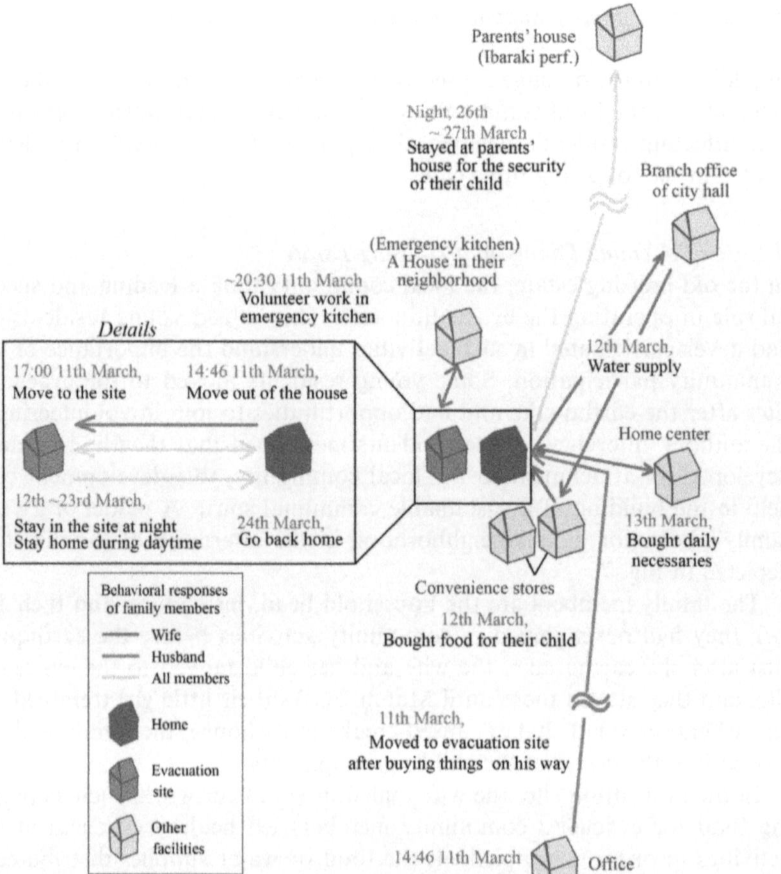

Fig. 7. A Case of a Young Family in Hitachi City After the Tohoku Pacific Earthquake. *Source:* Authors' Interview Survey.

evacuation sites, they preferred relying on their connections. Fig. 8 depicts their behavior.

The family's house completely collapsed in the tsunami; its members, who run a coastal inn, are the household head (a fisherman), his mother, and his grandmother. After the earthquake, his grandmother moved to a relative's house in another area.

After the tsunami struck, the mother immediately moved to the evacuation site only to find the place in a chaotic state. The household head tried

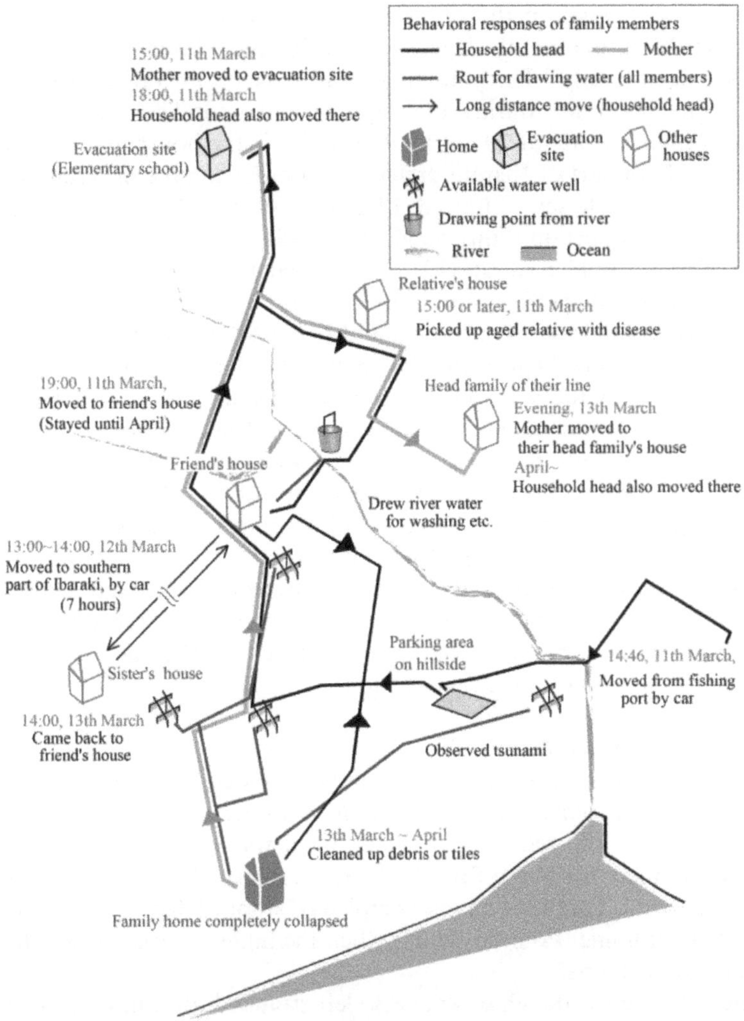

Fig. 8. A Case of a Tsunami-Affected Family in Hitachi City After the Tohoku Pacific Earthquake. *Source*: Authors' Interview Survey.

to return home, but he noticed that the tsunami was coming; he observed it from a parking area on a nearby hillside and saw his house collapse. At around 6 o'clock in the evening, family members met each other at the site and soon they moved to a nearby relative's house. They reached the head-family house (*honke*) in the evening; the household head's mother stayed there for about one month. It was natural for them to ask for the support for their head family.[6] The household head moved to his friend's house; he tried to fix his house during the daytime and went back to his friend's house at night for about 20 days. On March 12, the household head drove to the southern part of Ibaraki prefecture in order to take his grandmother there for safety. He says, "Indeed, I lost almost everything by the tsunami, but one most important thing, bond, remained."

The family could have stayed in the evacuation site, but they did not choose this option for several reasons: (1) The condition of the site was not perfect; some windows were broken and blankets, which should have been stockpiled, were not kept in the designated storehouse of their district. (2) The operation of the site was not successful, since the local community, whose members were mainly newcomers, did not take initiatives in operating it. (3) The family had lived in the neighborhoods for generations and had very strong local familial and friendship bonds. (4) Based on family ties, the family had its own channels of communication with neighbors, while newcomers were excluded from this system.

For the reasons mentioned above, the residents of coastal settlements tended to move out of the evacuation site. This finding suggests that not only its condition but also the characteristics of the local community affected the success of disaster operations.

A Case of a Stranded Family

Finally, we discuss the case of a family stranded after the earthquake (Fig. 9). It took four days until all family members returned home. When the earthquake occurred, the household head was in his Tokyo office and had to stay there for about four days. As the railroad service to the northern Kanto and Tohoku regions completely stopped for several days, he could not get home. Four days later when the highway was opened, he traveled home on a bus.

The wife and the daughter were also left stranded; they moved to several evacuation sites on their way home from the daughter's school. At the time of the earthquake, the wife was on her way to pick up her daughter at a high school in a neighboring city. Late at night of March 11, the wife could find the location of her daughter and went home.

As telephone services did not work for several hours, it was difficult for most people to know if their family members were safe. Most Japanese have been trained to go to a nearby evacuation site, usually public schools. This training makes it easier for people to find where their family members are in

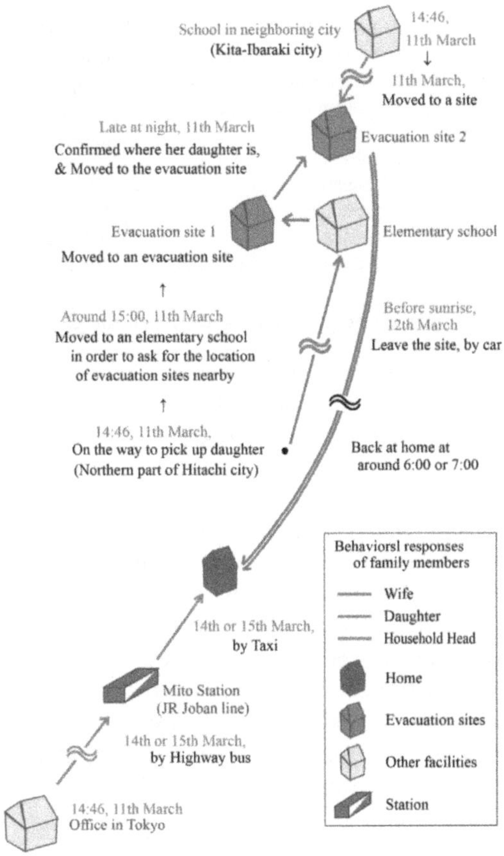

Fig. 9. A Case of a Stranded Family in Hitachi City After the Tohoku Pacific Earthquake. *Source:* Authors' Interview Survey.

such emergencies. In contrast to Edgington (2010), who notes that Japanese evacuation drills are more symbolic than practical, we demonstrate that the drills are fundamental in personal decision making during emergencies.

Changes of Neighborhood Perceptions Before and After the Disaster

Perception on Residential Environment

Fig. 10 indicates the changes of resident perceptions of neighborhood environments before and after the disaster. In most cases, residents

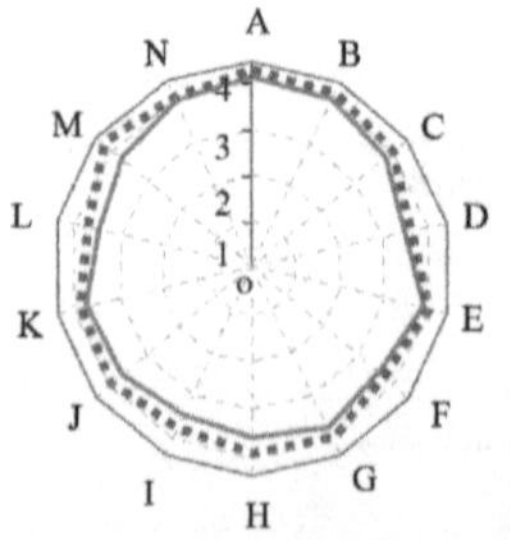

A: proximity to stations
B: proximity to offices
C: convenience of shopping
D: convenience of public transport
E: substantial educational facilities
F: proximity to medical facilities
G: substantial public facilities
H: maintenance of roads
I: maintenance of pedestrians
J: stability of land
K: security of neighborhoods
L: disaster-resistance system
M: public/private supplies (water, electricity etc.)
N: closeness with local community, friends, and relatives

··· Average score: Before the disaster — Average score: After the disaster

Fig. 10. Changes of the Residents' Perception to Their Residential Environment of Their Neighborhoods in Hitachi City (2012). *Note:* Adapted from Hashimoto et al. (2012).*Source:* Authors' Questionnaire Survey.

downgraded community scores after the disaster. Public and private supplies of energy, water, and sewage fell most in ranking, followed by the maintenance of roads and sidewalks. Surprisingly, many residents did not upgrade the scores of closeness with local communities, friends, and relatives. This stagnancy reflects diverse perceptions. Young residents, but not others, appear to have noticed the importance of their communities and upgraded these scores. If respondents were in the core regions of the disaster, the scores before and after it changed dramatically. As Heeren (1999) mentions, victims build up strong bonds with each other by sharing difficulties following a catastrophe; thus the scores of closeness with the local community are expected to rise.

Comments of Residents in a Coastal Settlement to the Disaster

Comments and requests presented by residents of a coastal settlement in Hitachi City were listed in Table 4. There were eight types of comments: (i) recovery of roads and sidewalks (22 answers), (ii) water supply (11), (iii) distribution of information (9), (iv) nuclear accident (8), (v) evacuation sites (12), (vi) Hitachi City and its officers (11), (vii) economic matters (8), and (viii) others and no comments (4 and 24 answers, respectively) which were excluded in the table. We can clearly see how people's lives are filled with uneasiness, confusion, and distrust to local and national governments.

Regarding such conditions, Hitachi City took several actions in 2012, for example, building emergency storehouses, conducting surveys on wells within the city, and modification of its hazard maps. Local communities,

Table 4. Comments on Disaster Responses by Residents in Coastal Settlement in Hitachi City (2011).

Types (Number of Answers)	Comments and Requests from Residents
Recovery of roads and sidewalks (22)	I worry about dangerous condition of roads and sidewalks in our city.
	After the earthquake, whenever trailers or big dumps pass the Route 245, we feel small quakes. Although it recovered apparently, it must be damaged underground.
Water supply (11)	We need a map of available well water for emergencies. People taught me locations of available well water and spring. It helped me so much.
	I hope the city subsidize its residents who own wells so that they can equip pumps under manual operation; local residents will be able to ask for water at power failure caused by natural disasters.
	All community centers, public facilities, and designated evacuation sites should sink wells in their lots.
Distribution of information (9)	I could not gain useful information about Ibaraki prefecture and Hitachi City. Most media information was limited to the Tohoku region.
	Besides emergency radios, we need emergency outside stereos.
	We could not understand any disaster information distributed by city's cars.
Nuclear accident (8)	"Fuhyo-higai" is serious. Please let people know that agricultural and aquacultural products in Hitachi are safe!
	As the Fukushima nuclear plant is relatively close to Hitachi, I want to know radiation level in Hitachi.
	We live 8 km away from Tokai-mura nuclear facilities. I wonder if these facilities consider measures for earthquake and tsunami. Urgent measures are necessary.
Evacuation sites (12)	We need safe evacuation sites!!
	Water, blankets, and foods .... We wanted them as much as we needed.
	Evacuation site was dark and cold, I preferred to stay at home.
	In addition to designated sites, we need storehouses for foods, water, blankets, and emergency power generators.
Hitachi City and its officers (11)	Hitachi City should distribute accurate information as much as possible. Although I went to a branch office to know when water and electricity services would recover, no officer could answer to it.
	Too late, too late.
	Officers were in the evacuation sites, but they could not do almost anything because the City did not give them orders.
Economic matters (8)	Housing recovery costs too much, and it's almost three times more expensive than usual.
	Daily goods became more expensive than before.

Note: 1. Free comments in authors' questionnaire surveys were used for the table. 2. In addition to the types of comments listed in the table, there were 4 comments classified as "Others" and 24 "No comments."

Source: Authors' questionnaire survey. Arranged in Hashimoto et al. (2012).

however, started to make their own disaster-prevention plans. Some core members of residents' associations mentioned that they would not believe the City, and they had to protect their residents by themselves. Conversation between the City officials and local communities are required in order to make sustainable and successful disaster management within the City. These comments can be the key to make better disaster-prevention plans and community-based urban planning for future.

These results may symbolize features of surrounding regions of disaster damages. People don't share the strong image that "we are victims of the disaster," they rather believe that only the people in the Tohoku region are the victims. Such image can affect people's perception of damages caused by the disaster. Although there are enough number of people whose houses are completely collapsed by tsunami in the same neighborhood, people may ignore this fact. In our interviews, some residents, who lost their house or forced to rebuild some parts of their house, had sighs over cruel opinions from other residents in the same neighborhood. They felt as if they were ignored by society, and that they had no right to talk about their hard time during and after disasters.[7]

DISCUSSION AND CONCLUSION

After the earthquake and tsunami, residents initially tried to go back to their homes. Depending on the damage, they remained in them, moved to relatives or friends' houses, or to neighborhood evacuation sites. The disruption of basic public services, transportation systems, and other damages limited the mobility of residents who had only the human and physical resources of their neighborhoods at their disposal. Therefore, the characteristics of local communities affected resident behaviors during and after the earthquake. Evacuation drills reinforced the operation of evacuation sites in a hillside community.

Shaw and Goda (2004) clearly explain that civil society is the base of a sustainable community. Leadership by residents' associations is crucial in the making of day-to-day and emergency decisions. Edgington (2010) emphasizes that cooperation between local authorities and residents (or residents' associations) was essential in Kobe's sustainable reconstruction plans. A human environment consists of more than buildings, roads, and parks; it also needs social networks, such as residents' associations or local small businesses (Kubo, Onozawa, Hashimoto, Hishinuma, & Matsui,

2010). Indeed, the leadership of local communities can diversify resident benefits, which become evident during natural disasters. Young Japanese do not join residents' associations or partake of community activities in urbanized areas. Although they may participate in NGOs or volunteer activities, they regard residents' associations as senior clubs. The earthquake and tsunami, however, gave young residents the chance to see that they were members of their neighborhoods. After the 1995 Great Hanshin-Awaji Earthquake, the involvement of local residents in policy making rapidly increased (Edgington, 2010). Although reconstruction from the 2011 Tohoku Pacific Earthquake has gone slowly, we clearly see greater local involvement and the power of grassroots activities into the rebuilding of local communities and in small-scale disaster preventions. Many university students and young people have joined volunteer activities in the disaster areas. Japanese society has started to build a sustainable civil society.

Community support activities and resident behavior indicate that the geography of the region or neighborhood matters in emergencies; both physical and human geography contribute to disaster reactions. As Kunugiza and Matsui (2005) note, our decision-making processes and area images are presented in a biased manner in mass media. Geography lessons contribute to better decision making in emergencies by offering regional knowledge to students, who will be very important actors in local communities in the near future. Those geography lessons will be discussed in our future studies.

NOTES

1. Wakabayashi (2009) notes that perceptional-behavioral geography enjoyed quite boom after the 1965 special meeting of the Association of American Geographers; its popularity peaked in the 1960s in the United Kingdom and in the late 1990s in Japan. The approach declined as geographers started to diversify their research interests. On the other hand, in the United States, perceptional-behavioral geography has gain great attention; now most studies with this method are published in the United States.
2. According to the Ibaraki prefectural Fisheries Administration Division, a total of 143 species of marine products and 163 kinds of processed marine products were examined to survey their levels of radioactive substances till May 28, 2013 in Ibaraki prefecture. Through these surveys, fishery and selling of marine products such as whitebaits (*Shirasu*), panther puffers (*Higan-Fugu*), gofers (*Usu-mebaru*), and so on started in the spring 2013 (http://www.pref.ibaraki.jp/nourin/gyosei/housyanou_jyouhou.html#2; accessed on May 31, 2013).
3. The JCO Nuclear Accident, which occurred in Tokai-mura village in September 1999, was a mishap in a uranium reprocessing facility operated by the

JCO, a nuclear fuel cycle company. A subsidiary of the Sumitomo Metal Mining Co., Ltd., it was founded in 1979. According to the International Atomic Energy Agency (IAEA), this accident was recorded as level four on the International Nuclear Event Scale (INES). Two people were killed and more than 660 were exposed to radiation. More than 160 residents living within a 350-m radius were asked to evacuate for five hours after the incident, and people in a 10-km radius were asked to stay indoors.

4. According to the report of the Japanese Reconstruction Agency, a total of 2,688 people were dead by the 2011 Tohoku Pacific earthquake and tsunami; 389 in Iwate prefecture, 862 in Miyagi prefecture, 2 in Yamagata prefecture, 1383 in Fukushima prefecture, 41 in Ibaraki prefecture, 4 in China prefecture, and 7 in other prefectures in the Kanto region (March 31, 2013).

5. Most Japanese suburban neighborhoods, which were built from the 1960s to 1980s, have witnessed a period of maturation and generational transition (Nakazawa, Sato, & Kawaguchi, 2008). They are now less popular than city-center condominiums or new bay-area housing developments (Hirayama, 2005) and face serious social problems, such as aging populations and worsening residential environments (Naganuma, Arai, & Esaki, 2006).

6. Before 1947, when the *ie-seido* (patriarchal family institution) was abolished, the relationships of head family and its collateral families, at the same time family head and other family members in a household, were regarded as those of masters and servants. Head families took responsibility for the lives of collateral families. Such relationships have now weakened in Japan as a whole, but they can still be seen in rural and fishing communities.

7. Regionally biased images and their demerits are examined in our next studies. A part of our results has already been presented in the 2012 annual conference of the Japan Studies Association of Canada (JSAC), which was held in Ottawa in October 2012.

ACKNOWLEDGMENTS

A part of this study is a revised and modified version of Hashimoto, Greger, Mashita, Yamamoto, and Kubo (2012). This study was financially supported by a JSPS Grant-in Aid for JSPS Research Fellow (Dr Tomoko Kubo, 23/7412) and a Grant-in Aid for Science Research by Tokyo Geographical Society (the representative of this project, Dr Keisuke Matsui).

REFERENCES

Alexander, D. (1993). *Natural disasters, Wellington after the quake: The challenges of rebuilding cities* (pp. 47–54). Wellington: Earthquake Commission and Center for Advanced Engineering.

Bolin, R., & Stanford, L. (1998). The Northridge earthquake: Community-based approaches to unmet recovery needs. *Disasters, 22*, 21–38.

Cabinet Office, Government of Japan. (2011). *Disaster management in Japan*. Director General for Disaster Management, Tokyo, Cabinet Office, Government of Japan.

Edgington, D. W. (2010). *Reconstructing KOBE: The geography of crisis and opportunity*. Vancouver: UBC Press.

Fujioka, H. (1996). Response to the earthquake disaster in Sake-brewing regions of Kobe City: Especially concerning the problem of water. *Geographical Review of Japan, 69A*, 547–558. (Written in Japanese with English abstract)

Fukutome, K. (1999). Relationship between damage and tenure of land and houses in the 1995 Hanshin-Awaji Great Earthquake: A case study in the district around Kamisawa Station, Kobe. *Geographical Review of Japan, 72A*, 668–690. (Written in Japanese with English abstract)

Ganzawa, Y., Kito, N., & Sadakata, N. (1995). Tsunami of the 1993 southwest off Hokkaido earthquake and refuge behavior of people: In the case of Taisei town in Hokkaido, Japan. *Earth Science (Chikyu Kagaku), 49*, 379–390. (Written in Japanese with English abstract)

Gold, J.R. (1980). *An introduction to behavioral geography*. Oxford: Oxford University Press.

Hashimoto, M., Greger, K., Mashita, M., Yamamoto, T., & Kubo, T. (2012). The roles of local community in Hitachi City, Ibaraki prefecture after the Tohoku Earthquake concerning damage -prevention plan and the operation of evacuation sites. *Annals of Human and Regional Geography, 34*, 111–136. (Written in Japanese)

Heeren, J. W. (1999). Emotional simultaneity and the construction of victim unity. *Symbolic Interaction, 22*(2), 163–179.

Hirayama, Y. (2000). Collapse and reconstruction: Housing recovery policy in Kobe after the Hanshin Great Earthquake. *Housing Studies, 15*, 111–128.

Hirayama, Y. (2005). Running hot and cold in the urban home-ownership market: The experience of Japan's major cities. *Journal of Housing and the Built Environment, 20*, 1–20.

Ishii, M., Yamazaki, K., Namai, S., Uchida, H., & Okazawa, S. (1996). Regional characteristics of human injury in the Hanshin-Awaji Great Earthquake Disaster: With special reference to severely damaged areas in Kobe city. *Geographical Review of Japan, 69A*, 559–578. (Written in Japanese with English abstract)

Kagawa, T. (1995). Damage to Toyonaka City caused by the earthquake in the Southern part of Hyogo prefecture. *Geographical Review of Japan, 68A*, 550–562. (Written in Japanese with English abstract)

Kanasaka, K. (1995). Sinsai-houdoi to Hisai-tiiki (Information and disaster-stricken region). *Geography, 40*(4), 66–71. (Written in Japanese)

Kates, R. W., & Pijawka, D. (1977). From rubble to monument: The pace of reconstruction. In E. J. Haas, R. W. Kates, & M. J. Bowden (Eds.), *Disaster and reconstruction* (pp. 1–23). Cambridge, MA: MIT Press.

Kubo, T., Onozawa, Y., Hashimoto, M., Hishinuma, Y., & Matsui, K. (2010). Mixed development in sustainability of suburban neighborhoods: The case of Narita New Town. *Geographical Review of Japan, 83*, 47–63.

Kunugiza, K., & Matsui, Y. (2005). Reconstruction of disaster reduction education with special reference to Tsunami disaster risk and the necessity of an earthquake resistant house. *Reports of University of Toyama (Toyama daigaku kenkyu ron syu), 8*, 57–74. (Written in Japanese)

Massard-Guilbaud, G. (2002). Introduction: The urban catastrophe – Challenge to the social, economic, and culture order of the city. In G. Massard-Guilbaud, H. L. Platt, & D. Schott (Eds.), *City and catastrophes: Coping with emergency in European history* (pp. 9–42). Frankfurt am Main: Peter Lang.

Mather, C., Karan, P. P., & Iijima, S. (1998). *Japanese landscape: Where land and culture Merge*. Lexington: University of Kentucky Press.

Mileti, D. (1999). *Disasters by design: A reassessment of natural hazards in the United States*. Washington, DC: Joseph Henry.

Murakami, S. (1996). Ready or not: How prepared is Japan for the next major quake? *Look Japan* (February), 10–12.

Naganuma, S., Arai, Y., & Esaki, Y. (2006). The aging population in the suburbs of Metropolitan Tokyo. *Japanese Journal of Human Geography*, *58*, 399–412. (Written in Japanese with English abstract)

Nakazawa, T., Sato, H., & Kawaguchi, T. (2008). First generation aging, second generation on the move: The process of generation transition in two suburban neighborhoods in the Tokyo Metropolitan area. *Japanese Journal of Human Geography*, *60*, 144–162. (Written in Japanese with English abstract)

Okamoto, K. (1998). History of behavioral geography and its future. *The Human Geography (Jinbun-Chiri)*, *50*, 23–42. (Written in Japanese with English abstract)

Ozerdem, A., & Jacoby, T. (2006). *Disaster management and civil society: Earthquake relief in Japan, Turkey, and India*. London: Tauris.

Pekkanen, R. (2006). *Japan's dual civil society*. Stanford, CA: Stanford University Press.

Schwartz, F. J., & Pharr, S. J. (2003). *The state of civil society in Japan*. Cambridge, MA: Cambridge University Press.

Shaw, R., & Goda, K. (2004). From disaster to sustainable civil society: The Kobe experience. *Disasters*, *28*, 16–40.

Takeda, J., Tamura, K., & Tatsuiki, S. (2003). Life recovery of 1995 Kobe Earthquake survivors in Nishinomiya City: A total-quality-management-based assessment of disadvantaged populations. *Natural Hazards*, *29*, 565–583.

Wakabayashi, Y. (2009). Progress in perceptual-behavioral geography in Japan: Retrospect and prospect. *Japanese Journal of Human Geography*, *61*, 266–281. (Written in Japanese with English abstract)

Wisner, B. (1998). Marginality and vulnerability: Why the homeless of Tokyo don't "count" in disaster preparations. *Applied Geography*, *18*, 25–33.

APPENDIX

Table A.1. Annexure: Major Natural Disasters and Disaster Management Laws in Japan (1945–2010).

Major Natural Disasters or Events in Japan		Disaster Management Laws Designated After Natural Disasters or Events	
Year	Natural disasters or events	Year	Disaster management laws or conditions
1946	Nankai Earthquake	1947	Disaster Relief Act
1945	Typhoon Makurazaki	1949	Flood Control Act
1947	Typhoon Catherine		
1948	Fukui Earthquake	1950	Enacting Standard Law
1959	Typhoon Ise-wan	1960	Soil Conservation and Flood Control Urgent Measure Act
		1961	Disaster Countermeasure Act
		1962	Act on special financial support to deal with extremely severe disaster
1961	Heavy snowfall	1962	Act on special measures for heavy snowfall areas
1964	Niigata Earthquake	1966	Act on earthquake insurance
1973	Mt. Sakura-jima Eruption	1973	Act on special measures for active volcanoes
1973	Mt. Asama Eruption		
1976	Seismological Society of Japan of Tokai Earthquake	1978	Act on special measures for large-scale earthquakes
		1980	Act on special measures for urgent earthquake countermeasure improvement projects in areas with intensified measures
1978	Miyagi-ken-oki Earthquake	1981	Amendment of Building Standard Law
1993	Southwest off Hokkaido Earthquake		Magnitude 7.8; a total of 201 people died
1995	Great Hanshin-Awaji Earthquake	1995	Act on special measures for earthquake disaster countermeasures
		1995	Act on promotion of the earthquake-proof retrofit of buildings
		1995	Amendment of Disaster Countermeasure Act
		1995	Amendment of act on special measures for large-scale earthquakes
		1996	Act on special measures for preservation of rights and profits of the victims of specified disasters
		1997	Act on promotion of disaster resilience improvement in densely inhabited areas

Table A.1. (*Continued*)

Major Natural Disasters or Events in Japan		Disaster Management Laws Designated After Natural Disasters or Events	
Year	Natural disasters or events	Year	Disaster management laws or conditions
		1998	Act on support for livelihood recovery of disaster victims
1999	JCO nuclear accident	1999	Act on special measures for nuclear accident
1999	Torrential rains in Hiroshima	2000	Act on promotion of sediment disaster countermeasures for sediment disaster-prone areas
2000	Torrential rains in the Tokai region	2001	Amendment of Flood Control Act
		2003	Special Urban River Inundation Countermeasures Act
2004	Niigata-Fukushima torrential rains, etc.	2005	Amendment of Flood Control Act
		2005	Amendment of act on promotion of sediment disaster countermeasures for sediment disaster-prone areas
2004	Niigata-Chuetsu Earthquake	2005	Amendment of act on promotion of earthquake-proof retrofit of buildings
		2006	Amendment of act on the regulation of residential land development

Source: Arranged by Cabinet Office, Government of Japan (2011).

CHAPTER 3

FILLING THE GAPS FROM THE CHRISTCHURCH EARTHQUAKES 2010–2013: GREENING THE RUBBLE AND THE MT PLEASANT COMMUNITY RESPONSE PLAN AS TWO LOCAL INITIATIVES

Roy Montgomery

ABSTRACT

Between September 4, 2010 and mid-2013 a severe earthquake sequence struck Christchurch, the second largest city in New Zealand, causing multiple fatalities and the destruction of much of the central business district. Large areas of suburban residential housing were condemned with the prospect that entire neighbourhoods would be abandoned for several decades if not permanently. The recovery and rebuilding process was immediately placed high on central and local government agendas since Christchurch and the surrounding Canterbury region were and continue to be seen as crucial to the security and stability of the national economy. Programmes for recovery developed initially relied principally on one-off

Risks and Conflicts: Local Responses to Natural Disasters
Community, Environment and Disaster Risk Management, Volume 14, 43–78

ISSN: 2040-7262/doi:10.1108/S2040-7262(2013)0000014009

funding packages and strategies from central government, local government recovery plans and the settlement of commercial insurance claims. There remains, however, the spectre of Christchurch as a city of demolition sites and vacant lots for the best part of a decade if not longer. Furthermore, although local and national Civil Defence and Emergency Management systems were activated during the most severe seismic events the response operations did not always reach those in need as promptly as was expected. Residents in a number of communities and neighbourhoods are now conscious that when disaster strikes they are still likely to have to fend for themselves. This chapter documents and evaluates two specific "gap-filling" responses to the Christchurch earthquakes over a three-year period. The first response considered is a community-based project called "Greening the Rubble" which took root in October 2010 as the prospect of a central city of vacant lots and car parks worried a number of volunteers into action to temporarily cheer up empty public and private sites with pocket parks, native plant displays and cultural interventions. The second initiative scrutinised, the "Mt Pleasant Community Response Plan 2012–2013," is one of the first community-based emergency response plans to emerge that has sought to complement official civil defence planning arrangements. Both responses are discussed in detail in the context of constantly changing and evolving hazardscapes and socio-economic and political conditions.

Keywords: Community response; urban ecology; resilience; new institutions

INTRODUCTION

On September 4, 2010 at 4:35 a.m. a magnitude 7.1 earthquake occurred in the Canterbury region in the South Island of New Zealand. The epicentre was located under a sparsely populated rural farming area known as Greendale which is close to Darfield, a rural service town of some 1,400 people. Due in part to the earthquake's relative shallowness, considerable damage was done to buildings and infrastructure in adjoining towns and urban areas, including the city of Christchurch some 40 kilometres away. There were no fatalities as a result of the "Darfield earthquake" but across the region the damage to unreinforced civic buildings and churches of the late Victorian and Edwardian eras was such that it was clear that many structures across would have to be rebuilt if not completely demolished.

In Christchurch, often presented and perceived as a very English city with a well-regarded 19th and early 20th century architectural heritage, there was great apprehension that the city's identity was at risk. The city centre or central business district (CBD) had already been hollowed out by suburban retail and service competition from the 1980s onwards. People feared that this atmosphere of desertion would be further blighted by the standard fallback use for urban "greyfield" vacant sites – car parking lots.

Furthermore, the characteristic hydrology and geology of the Canterbury Plains – it is a broad, relatively low-lying floodplain with multiple layers of aquifers – meant that liquefaction under severe shaking was likely to be a major effect. This was instantly apparent in the riverine and estuarial parts of suburban housing areas. Such were the liquefaction effects that many private dwellings partially collapsed and the land under them subsided severely. Extensive land remediation would have to be undertaken before reconstruction could take place. In effect, large areas of suburban residential housing were condemned with the prospect that substantial parts of particular neighbourhoods would be abandoned for several years as the land remediation technology was trialled and perfected.

On top of this, and as seismic experts had predicted, an aftershock sequence was triggered and this generated further damaging events, notably one on 26 December 2010. However, while these "Canterbury Earthquakes," as they were termed, caused considerable trauma and disruption to daily life, the overall mood among Cantabrians whose homes were not significantly damaged was that "business as usual" was possible within a relatively short period of time, i.e. months rather than years. Indeed, the Mayor of Christchurch was quoted as saying so within seven days of the first earthquake.

On 22 February 2011 at 12:51 p.m. any complacency evaporated instantly as a 6.3 magnitude aftershock struck along a fault previously unidentified, the Port Hills fault, which lay within 10 kilometres of the centre of the city. This new "Christchurch earthquake" as it was labelled (and there is still conjecture as to whether it is an aftershock or separate earthquake since it derives from a separate fault to the Greendale fault) caused multi-storey buildings in the CBD to collapse and rocks to fall in hillside suburbs causing 185 fatalities, 164 major trauma injuries and minor injuries to several thousand people. In geological and seismological terms the aftershock was classified as a strike–slip event with oblique motion and reverse thrust. To scientists and lay people alike this was the worst kind of combination possible: horizontal and vertical ground movement with extreme upward acceleration.[1]

The liquefaction process returned to the areas struck by the previous earthquake and aftershocks and extended the amount of land choked by sand, mud and water by vast proportions. Some areas were left as much as 30 centimetres lower than they had been before the earthquakes. Talk of land remediation halted abruptly. It was anticipated that more than 5,000 homes in the low-lying, low-income eastern suburbs might have to be abandoned indefinitely.[2] Furthermore, while the more affluent hill suburbs had remained relatively unaffected by the Darfield quake, the February event caused major devastation to houses on the eastern flanks of the Port Hills (Fig. 1) and left parts of the Port Hills some 40 centimetres higher than they had been prior to the earthquakes. In this sense the aftershock acted as a great socio-economic "leveller" producing some otherwise unlikely shared experiences, e.g. improvised backyard accommodation and toilets.

Naturally, the degree of official emergency management and government response was correspondingly much greater than had been the case in September and a cascade of local, national and international crisis response

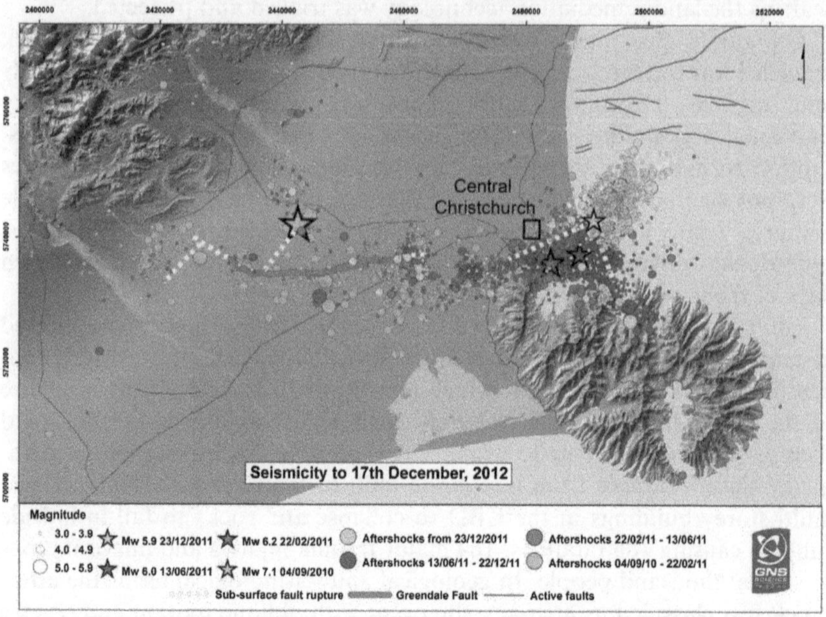

Fig. 1. Geological and Nuclear Sciences Crown Research Institute Map Showing Earthquake and Aftershocks Between 4 September 2010 and 17 December 2012. *Source*: Image Courtesy Geological and Nuclear Sciences, Wellington, New Zealand.

interventions quickly followed. Urban Search and Rescue (USAR) teams, Disaster Victim Identification (DVI) teams and Red Cross personnel from many countries arrived within two days. Yet both the September and February events created social resilience gaps, ruptures and opportunities that were unplanned for and which could not be addressed through existing institutional and social arrangements. Virtually overnight from February 22 the city lost nearly 10,000 permanent residents or 2.4% of its population, and this loss is not expected to be recovered for at least five years.

This chapter documents and evaluates two specific "gap-filling" responses to the Christchurch earthquakes over a three-year period. The first response considered is a community-based project called "Greening the Rubble" which took root in October 2010 as the prospect of a central city of vacant lots, and car parks worried a number of volunteers into action to temporarily cheer up empty public and private sites with pocket parks, native plant displays and cultural interventions. The second initiative described and examined is the "Mt Pleasant Community Response Plan 2012–2013" produced during the 2011 and 2012 period by a local community group working under the auspices of a residents association. It is one of the first community-based emergency response plans in New Zealand created without direct input or funding from official civil defence agencies. These local responses are discussed in the context of constantly changing and evolving hazards capes and in light of turbulent socio-economic and political conditions. The chapter concludes by referring back to current thinking in the emergency management, natural hazard and disaster response literature and posits factors to be considered for local responses to natural disasters in future settings.

METHODOLOGY

In simple terms I provide a narrative on two community projects with which I have been closely involved over the past three years. I have been on the interim steering groups and project advisory groups for Greening the Rubble since October 2010, and since January 2013 have been an official Trustee. I joined the Mt Pleasant Community Response Group in October 2011 and drafted most of the content of the first Mt Pleasant Community Response Plan.

In more academic terms the research described here can be characterised as a combination of ethnographic, participant observation, participatory action research and case study techniques informed by concepts from the literature referred to above. The research is further "grounded" or contextually situated to the extent that as an individual I experienced the initial earthquake and the major aftershocks together with the 11,000 aftershocks that have followed. I also face some of the personal issues faced by other project participants such as professional working environment disruptions, interruptions to school routines for children and a protracted wait for repairs to the residential property in which I live.[3]

The principal source materials for this research comprise a mixture of archived meeting notes, working documents, anecdotal and unrecorded conversations and personal observations and reflections. I have attended most of the scheduled meetings of both groups and participated in a number of public and outdoor events and have been involved in site construction and deconstruction work in the case of Greening the Rubble. Despite a putative degree of research triangulation that follows from these three varying experiential dimensions of a natural disaster there is of course an inevitable subjectivity against which one can claim only so much critical distance.

NEW ZEALAND IN CONTEXT

In broad terms New Zealand is a "young" country both geologically and culturally. Rocks dating back some 500 million years associated with the Gondwana landmass or supercontinent have been found and some 85 million years ago the archipelago recognised today as Aotearoa or New Zealand broke away from this large landmass. The alternating hill and plateau topography of the North Island and the mountainous character of much of the South Island are subject to significant volcanic and plate tectonic activity respectively. These features provide an obvious sign of New Zealand's location on the boundaries of the so-called Pacific Rim or ring-of-fire where the Australian Plate and Pacific Plate intersect. The land area of New Zealand is approximately 268,021 square kilometres or about 75% of the area of Germany to place this in a northern hemisphere context.[4]

Human settlement is of course much more recent. While debates continue as to the when and how of Polynesian migration (several "waves" over centuries or a condensed pulse?) it is estimated that the main period was AD 1280–1300. Although European colonisation began in the early 1800s planned settlement dates from the mid-19th century. At the present the population of New Zealand is approximately 4,430,400. Again, for the sake of comparison this is roughly 5.4% of the population of Germany. Population densities are 16.59 people per square kilometre and 234.35 people per square kilometre respectively. Prior to the earthquakes Christchurch was the second largest city with a population of 386,100 in an area of approximately 452 square kilometres compared to Cologne, the fourth largest in Germany with a population of 1,007,119 contained within an area of 405 square kilometres.

By such measures New Zealand is a sparsely populated country and although population distribution mirrors global patterns insofar as more than three quarters of its people live in urban areas these cities tend to be low in density. To that extent New Zealand resembles much of Australia and western parts of the United States and the notion of the "rugged pioneer" is deeply ingrained in New Zealand culture. Self-sufficiency and an ability to improvise, the metaphor used here, borrowed from the agricultural history of settlement, is "the number eight wire approach" which means the ability to do running repairs on anything provided that some fencing wire is at hand. This "do it yourself" attitude or norm helps to explain what in the past accounted for high degrees of volunteerism in New Zealand society. As an example, there are 1,707 paid and 8,000 unpaid volunteer firefighters to cover all of New Zealand. A high percentage of volunteer firefighters relative to those who are paid is not unusual in Westernised countries but at 82% the New Zealand figure is significantly higher than the United States where volunteers made up 69% of the force in 2011.[5]

It is true that volunteerism and conventional sport and cultural club membership have eroded here as elsewhere over the past 50 years as social ties have changed. However, the base level of social capital, social networks and the capacity for face-to-face constructive social interaction is still perceived as very high despite the underlying and sometimes counter-veiling principle of privacy of the individual that is as strong in New Zealand as it is in Europe or North America. Overall it is fair to say that New Zealanders aspire to being seen as belonging to a multi-cultural and post-colonial society on a par with other liberal democratic societies. New Zealand is still widely regarded as one of the least corrupt countries in the world. Yet there is a sense that it can at present only just manage to maintain the standard of living for a "developed" economy and the economic and social fabric, particularly since the so-called "global financial crisis" of the late 2000s appears to be stretched very thin, sometimes prompting calls for union or integration with its larger neighbour Australia.

THE RESPONSE AND RECOVERY "MILIEU" FOLLOWING THE EARTHQUAKES

The Institutional Framework for Response

Like most nation states New Zealand has a comprehensive emergency management system in place to deal with natural hazard events and natural disasters. The Ministry of Civil Defence and Emergency Management (MCDEM) which is located within the Department of Internal Affairs is the lead agency and it is based in Wellington, the capital of New Zealand. Head office staffing comprises some 50 personnel and the organisation relies on a highly devolved delivery structure that conforms to, and draws upon, local and regional government jurisdictions, facilities and resources. The devolved units are known as Civil Defence and Emergency Management (CDEM) Groups. In the event of small-scale or locally contained events the relevant CDEM Group may act in the principal response and planning role drawing upon outside and head office resources as and when necessary. In the event of a significant disaster an Officials Domestic and External Security Committee (ODESC), administered by the Office of Prime Minister and Cabinet, is convened and lead by the Director of CDEM Group. This brings in the chief executives and other staff from entities such as the Ministry of Defence, the Police Department and the New Zealand Fire Service.

In Canterbury the CDEM Group is made up of 10 district councils and 1 regional council. The full-time Civil Defence paid staff at the Group level number less than 20 people. However, anyone working as a local government employee can, by statutory provision, be directed immediately into Civil Defence roles in the event of a local or national emergency. They will tend to be deployed to the main Emergency Operations Centre and local authority Service Centres, i.e. Council offices and Area Welfare Centres (usually designated as the larger urban or suburban publically funded high schools). However, a great part of the staffing of the Area Welfare Centres and the entire staffing of the smaller Sector Posts (typically primary schools and local community halls) relies on trained Civil Defence volunteers of which there are approximately 750 although many of these may be inactive at any given time.

The Response After 4 September 2010

A local state of civil emergency was declared in Canterbury after the September earthquake and the organisational structure as outlined above was activated in key locations close to where the main damage occurred. Civil Defence and local authority staff moved into key emergency operations and welfare centre roles in certain suburbs and outlying townships. Parts of the central city of Christchurch were cordoned off, as were some of the older suburban shopping centres and main street areas. Since no casualties had occurred, attention and speculation quickly turned to what structures would need to be demolished and how long it would take to rebuild perhaps dozens of commercial and civic buildings. The state of emergency was in place for only 12 days and it did not see large movements of people and supplies from other parts of the country and overseas into the affected areas.

Instead, the main flurry of activity was in Wellington. Central government moved quickly to pass legislation and create a new legal entity when it was realised that while the Civil Defence Emergency Management Act (2002) dealt reasonably well with immediate threat-to-life and welfare response needs it did not have sufficient authority or direction with regard to recovery and reconstruction. A Minister for the Christchurch Earthquake Recovery was appointed two days after the main event. Immediately prior to the lifting of the state of emergency legislation was drafted and the Canterbury Earthquake Response and Recovery Act was passed under urgency. The Canterbury Earthquake Recovery Commission was established to advise Crown Ministers on matters such as Orders in Council that might be needed to amend any other legislation that threatened to impede recovery activity.

For the most part Cantabrians affected by the earthquake found themselves preoccupied with getting repairs to their houses done, finding temporary accommodation, pondering loss of employment where businesses had been shut down and making claims under the state-funded Earthquake Commission and their own private insurers in cases where it appeared likely that the state-funded cover did not extend to the damage that they had experienced. There were complaints in some communities about Civil Defence and local authority infrastructure maintenance teams failing "to turn up" to a number of the designated Civil Defence Sector Posts as expected on the day of the earthquake. Television images of tired residents battling to clear their properties and neighbourhood streets of silt prompted spontaneous public action. Novel and initially unincorporated entities such as the Student Volunteer Army, the Farmy Army, and a multi-neighbourhood lobby group called the Canterbury Earthquake Recovery Network (CanCERN) to speak on behalf of damage-affected neighbourhoods sprang into existence within days of the event.[6] Concerns were expressed about what could be relied upon by way of planned emergency management response should something like this happen again.

By and large, however, the public mood was one of worrying about whether damaged heritage buildings were going to come down and how many demolition sites claim and, as noted in the introductory paragraph, the degree to which the city would be scarred by unsightly vacant lots and unnecessary car parks. It was in this context, recovery in the public or urban domain, that entities and initiatives such as Greening the Rubble, Rebuild Christchurch and an exhibition and discussion series by the New Zealand Institute of Architects called "Before After" were conceived and launched.

The Response After 22 February 2011

The February 2011 event triggered a national state of emergency which lasted for two months. The legal and institutional frameworks were changed yet again to give even greater emergency or executive powers. Christchurch and the surrounding Canterbury agricultural region were seen as crucial to the security and stability of the national economy. Much of rural Canterbury had only recently been converted from dry land pastoral farming to intensive dairy operations, and milk products were the principal source of New Zealand's export revenue. The Canterbury Earthquake Recovery Authority (CERA) was established under the Canterbury Earthquake Recovery Act by mid-April 2011. Later on a Central City Development Unit (CCDU) was created by CERA which effectively took control of the planning for reconstruction of the central city. CCDU produced its own plan and to many this appeared to usurp the recovery planning activities of the Christchurch City Council (CCC).

Furthermore, because in the February aftershock people had died, including many foreign nationals, in collapsed buildings, on city and suburban streets and in hillside areas, a Royal Commission of Inquiry was established and much greater scrutiny was applied to the emergency management agencies to the extent that almost all of them embarked upon official reviews of their operations to accompany the coroner's inquests and commissions of inquiry. Criticism of the Civil Defence "failure-to-show" response in suburban areas was repeated and amplified within some communities. This was particularly so in communities situated on the Port Hills. Residents there claimed that they were effectively abandoned to fend for themselves for several days and were only brought to the attention of the authorities by media helicopter flyovers or second-hand reports. Out of this dissatisfaction and surprise emerged a resolve in a number of city-edge communities such as Sumner, Redcliffs, Mt Pleasant and Lyttelton to create community response plans with or without the official sanction of Civil Defence or local authorities.

The overall milieu after February and throughout the remainder of 2011, aside from the continuing aftershock sequence which had everyone on edge, was thus a confusing mélange of relatively routine emergency assistance actions, e.g. Red Cross grants to families, one-off funding packages for small businesses, local government recovery plans, disputes about commercial and private insurance liability, jurisdictions, emergency powers, orders in council, newly created government institutions, and who was in charge, i.e. CERA, CCC, EQC or the insurance companies. There was also the considerable change of scale of damage with which to contend. The hundred or so damaged central city sites of September jumped to thousands from February in an instant. From the rubble and confusion emerged Greening the Rubble and the Mt Pleasant Community Response Plan.

GREENING THE RUBBLE

The Catalyst for Greening the Rubble

The deliberate ecological or botanical reclamation of neglected urban spaces is by no means a recent phenomenon and the pedigree of "urban ecology" has been discussed at some length (Montgomery, 2012). Thought had already been given to how "messy ecosystems," to borrow from Nassauer, could be brought gently to the conservative "garden city" of Christchurch.[7] This relatively esoteric debate mattered little to the ordinary citizens but within days of the September event people started putting pen to paper in the local media about what to do with the ruins. The first public comment on this was provided by the Canterbury Regional Biodiversity Co-ordinator Dr Wayne McCallum. In a letter to the Editor of *The Press* in late September, he coined the phrase "greening the rubble." McCallum used these words to describe both the symbolic and material redemption of damaged sites by the temporary installation of native plants and/or cultural events. Days later McCallum put out a web- and email-based call to attract potential willing helpers to make at least one project happen.

The first Greening the Rubble (GtR henceforth) "expressions of interest" meeting took place on 6 October 2010. Encouragingly, local Members of Parliament voiced their support for the idea and perhaps most crucially the idea found favour among councillors and relevant staff at CCC. A more formal meeting took place on October 12 from which a steering group of approximately 10 members was formed.[8] Over the next several weeks both the day-to-day project-focussed steering group and a longer-term strategy group met to discuss how to get things moving quickly.

The Initial Mode of Working

The general pattern that was established involved meetings averaging 10–20 people every 2–3 weeks from October 2010 to February 2011. The working principle was to fill the gap of rebuilding with temporary parks, community gardens and interesting green/cultural spaces with an emphasis on native plants appropriate to the area (Fig. 2). It was axiomatic that GtR

Fig. 2. Victoria Green: First Greening the Rubble Site December 2011 Featuring Wildflowers and Gabion Baskets Filled with Broken Bricks. *Source:* Greening the Rubble Archive.

activities should not compromise the rights of land owners. For each project GtR would sign a "licence to occupy" agreement and would be ready to vacate any site within four weeks. Volunteers would not excavate, set foundations or disturb the archaeology of historic sites. They would work with what was already in place or on the thin film of sealing materials that follow demolition and clearance. They would plan for the relocation of most, if not all, materials from one project to the next.

In terms of designs a temporary park with seating and grassed areas was an obvious way of dealing with a site (and this was the case with the very first GtR project). Yet the group was strongly driven by a desire to reflect what the local community, the owner and key former tenants or occupants wanted in that space. Some sites lent themselves well to park-making while others were "edgier" and needed to fit with the gritty or built-up surrounds. As an example, if the local community aspiration was for a community garden and the owner agreed, then the group tried to help make that happen. Most crucially, the core members relied on a loose network of volunteers to prepare, maintain, deconstruct and transport materials. We also aimed each time to find a different set of local volunteers to adopt any given site so that (a) it was locally "owned" and (b) our 50 or so regular participants did not burn out from serial project fatigue.

The Unanticipated Change

Prior to the February 2011 aftershock the new group saw its general scope as aiming for permission to work on 10 sites out of a possible 50–100 new vacant lots spread across Christchurch and in the neighbouring township of Kaiapoi to the north. The February event changed the situation considerably. While before virtually every GtR participant had avoided the direct effects of the earthquake, now many of them had compromised home and work circumstances and some were forced to leave the city. Official projections for the number of demolitions within the central city quickly topped 1,000 but these were only the multi-storey notables. The more likely and politically and economically more sensitive estimate was somewhere between 3 and 5,000 vacant commercial sites over an undefined period (politically sensitive to admit precise level of withdrawal from central city). This was extremely daunting. Yet somehow the group reconvened during March 2011 and took stock of what could be done to continue (Figs. 3 and 4). Sites that were cut-off by

Fig. 3. Damaged Shopping Block, Colombo Street, Sydenham, October 2010 After First Earthquake (Entire Block Demolished After February 2011 Aftershock). *Source:* Roy Montgomery.

Fig. 4. Greening the Rubble Installations Partially Completed on Colombo Street, Sydenham Former Shopping Block Shops Sites Early 2012. *Source:* Roy Montgomery.

safety cordons were left idle and some of those mooted projects that had seemed priorities were put aside while the new landscape was assessed.

Gradually, the steering group of approximately 12 individuals distilled into a smaller core of 6 plus professional advisors and designers (e.g. landscape architects) working pro bono, 2 part-time paid project workers and a pool of 100 volunteers. The question of status as an official charitable entity arose at the end of 2011 and throughout 2012 this option was pursued and Trust status was achieved in February 2013. CCC recognised the modest but meaningful work of the group by providing funding support in 2011 and 2012. For 2012–2013 an average of 10 new projects was expected by CCC in return for a $100,000 grant. Such funding allowed for the paid employment of a "project worker" or co-ordinator and the group could sometimes expect the assistance of City Care (the equivalent of a city works department) where this was deemed appropriate. A number of businesses and charitable trusts donated plants and materials and a number of large organisations and corporates made their staff available for public good works' days.

Bedding In

While it was challenging to co-ordinate all these inputs the support on the ground the response was extremely encouraging and the willingness of people to "muck in" was remarkable and it suggested that a solid base of social capital had survived the earthquakes. As more projects evolve people seem quite happy to lend their labours without necessarily understanding the full vision for a particular site or for GtR in general. GtR benefitted from word of mouth, the use of web-based communication and social media such as Facebook. News media coverage was consistently positive. A highlight was a special audience with *Grand Designs* television presenter and sustainable architecture and urban design advocate Kevin McCloud on 25 October 2011 during his first flying visit to Christchurch. Official recognition came by way of a Christchurch Civic Trust Award on 16 November 2011. These awards typically recognise significant contributions to civic life by individuals over a long period of time but in 2011 they were specially calibrated to earthquake recovery initiatives, individual and collective. Furthermore, one of the initial aims of GtR was to act as a conduit for independent but like-minded grassroots initiatives and this now appears to be bearing fruit with the emergence of such projects as Greening Spaces and Silty Bricks.[9] It is also worth noting that many of the people driving GtR have a long history in environmental advocacy and project work and GtR has provided an opportunity to deal with new constituencies, both public and private.

Learning by Doing

By early 2013 GtR had carried out approximately 15 projects and most are still in operation at the time of writing.[10] However, as time has passed it has become increasingly important to try to take stock of this spontaneous local response to a natural disaster. The lessons learned to date by those overseeing GtR revolve around specific themes and they can be summarised as follows.

People, Psychology and Perception Factors

- Organisers, designers and volunteers typically have their own personal earthquake challenges to deal with and the city is still very much in the middle of the event – decisions and rulings by various authorities can retard or inhibit the project efforts in some way as people drop out temporarily or permanently to deal with their own issues.
- Ensure that site owners are aware that they have signed obligations not to modify sites without notice and potentially jeopardise human health – they will sometimes forget that they have invited teams of strangers on to their property.
- Don't take the initially agreed site exit date too seriously – the pattern to date has been site owner requests to stay longer than was initially planned.
- Public and local authority perceptions and expectations about underutilised city sites often revolve around "tidiness" – wild or native nature is often "messy" at first glance and the gaining of acceptance for, and management of, "re-wilding" sites will pose considerable challenges.

A New Institutional Context

- Don't take new or existing governance and institutional frameworks at any given time for granted as they can change at very short notice.
- The GtR initiative is but one of a raft of novel quake entities that have appeared and which are likely to come and go over the next few years and there is the threat of dilution of both volunteer effort and media attention and the risk of confusion as to who does what.[11]
- If the lifespan of the Trust reaches the five-year mark it will face the "green guerrilla" dilemma of permanence versus contingency, and if it becomes part of "the establishment," will the temporary use philosophy run too thin to be credible?

Changing Circumstances Long After the Disaster Event

- Be prepared for sectors of the city to be shut down or opened up at short notice – this may stop activities already initiated in their tracks or put pressure on the Trust to finish work hurriedly.
- The destruction of the CBD has been so comprehensive that it diverts attention from the damaged outer suburbs and puts pressure on the Trust to focus its work on the central city at the expense of neighbourhoods.

Design Factors

- To maintain or increase its output on a modest budget and without over-exploiting volunteers or donors there is a need to move to standard designs that balance the need for professionalism and safety with community spontaneity.[12]
- It is important to standardise design elements and make them more modular and portable without becoming predictable and too formulaic (e.g. how to get the signature brick fragment gabion baskets from one place to another without having to empty them each time).
- On-site signage and "branding" are as important as social media for publicity and community understanding.[13]
- At some point the issue of prioritising locally sourced and ecologically appropriate native plants rather than any quick fix of indigenous and exotic plants will have to be addressed.

GtR as a vehicle for community recovery rather than emergency response and some of the points above are considered in more detail towards the end of this chapter. The next section deals with a specific emergency response local initiative.

THE MT PLEASANT COMMUNITY RESPONSE

Up until February 2011 Mt Pleasant was regarded as a relatively affluent, if sleepy, hill suburb of Christchurch numbering some 4,000 people and 1,600 dwellings.[14] It is situated on the eastern fringe of the city and overlooks a large estuary with commanding views of the Canterbury Plains, the Southern Alps and the Pacific Ocean. As a suburb it is a relatively recent development with most of the housing stock dating from the 1940s to the 1980s. Many of those who built or purchased homes in earlier decades had raised families to maturity and as retirees in the new millennium expected to age quietly in place. Two churches and a community centre had been built during the 1940s and 1950s and in the post-war period local residents could enjoy handily located yacht clubs, bowling clubs, and water sports associations. Two small shopping centres were established in the 1960s along with a medical centre. There were no factories, large commercial businesses, public institutions or other significant employers within the community apart from the local primary school. Apart from the medical centre there were no emergency services such as a fire or police station in Mt Pleasant.

The shopping centres had to compete with large suburban strip malls or shopping complexes and although a café and a restaurant had been set up by the turn of the 21st century there was little doubt that for outsiders Mt Pleasant was simply a location rather than a destination. The primary school, rated in 2011 as "decile" 10 on a 1–10 socioeconomic scale where 1 in effect means that the children's parents are very poor and 10 means they are very wealthy, had been rebuilt and expanded in the 1960s. This provided the only real growth indicator for the community during the early 2000s as younger adult couples and families started to colonise the housing stock left by retirees who had moved on. These younger families tended to be only loosely affiliated with existing community institutions and to a certain extent the social capital profile appeared to mirror that described by Robert Putnam in his lament over the loss of community in North America (Putnam, 2000).

Earthquake Damage on Mt Pleasant

In general terms the September earthquake caused more damage to pantry goods, crockery and unsecured chattels in Mt Pleasant than it did to dwellings and land. Most eyes and sentiments were directed towards the low-lying liquefaction-hit seaside and riverside communities. The February aftershock was dramatically different. Parts of Mt Pleasant lay only 1,000 metres from the epicentre and much of the suburb rode the edge of the Port Hills fault itself. Along with the neighbouring Heathcote Valley Mt Pleasant experienced the greatest vertical ground acceleration ever recorded in an earthquake. The land on Mt Pleasant is now on average 40 centimetres higher above sea level since the February event.

Many of the post-war brick veneer and concrete block wall and cement tiled-roof houses and modern structures were literally shaken to pieces although surprisingly no one died. Currently, the house occupancy rate is approximately 40% of its previous figure although there are no official statistics or records to show how many residents have been forced temporarily or permanently from their homes. Apart from the suburbs that have been entirely written off through government red-zoning of land it is now the most deserted suburb in Christchurch eclipsing the communities of Sumner, Redcliffs and Lyttelton, although the latter three enjoyed great international and local media attention in the days and months immediately following the February aftershock.

Improvised Response

On the day of the aftershock residents on the hill found themselves cut off from the city and other suburbs because the connecting roads on the flat were so badly damaged, inundated by mud and water or choked with gridlocked traffic that any notional emergency collecting points such as the high school welfare centre were rendered inaccessible. Teachers, parents and administrators at the local primary school, which was still in session at the time of the aftershock, and staff at the medical centre found themselves coping with a flow of displaced, disoriented and injured residents. Many assumed, understandably, that help would be found there or would soon arrive since there was a sign on the street outside the school identifying it as a Civil Defence Sector Post.

It quickly became apparent that organised help was not going to arrive any time soon. The minutes of the first Mt Pleasant Community Resilience Meeting of 1 September 2011, at which two Civil Defence staff were present, provide a narrative of how things unfolded:

> On February 22nd, Mt Pleasant School became an unofficial Sector Post. The school's expectation was that someone would arrive and lead the Civil Defence response locally and in the meantime teams were organized to retrieve gear, put up tents, dig toilets and arrange food. People from the local community started to arrive with injuries, illnesses, a young baby, and tour bus driver had directed tourists to the school also. Local residents started to report in about elderly people they were concerned about and others started to arrive for information. The Police arrived and told the school not to expect help. The Army arrived and gave the thumbs up and indicated people should arrive with soup. Heather McQuillan [Deputy School Principal] advised Sumner Fire/Police and CD looking [sic] person of school set up and felt confident this message would be passed on. Martin Anderson [Chair of School Board of Trustees] was unable to make contact with anyone on the list of Civil Defence phone numbers. Despite helicopters constantly passing overhead, and local emergency services being aware of the situation, no one at the Civil Defence Base appeared to be aware there were 300 people camping at the school until several days later.

> Fortunately there were parents of students who arrived that are Doctors, and others with connections to water supplies and "portaloos."[15] A rain water collection system was set up to collect rain.[16] Parents evacuating the area over the coming days also dropped off excess food supplies.[17]
>
> Civil Defence became aware of the set up at Mt Pleasant school when a parent went to the Burnside service centre. After that there were several phone calls a day offering support.

The sense of deep indignation and disillusionment may not be apparent from this account, but similar stories echoed across the city and in media reports it was clear that people were angry at what they thought was neglect from the Civil Defence assistance delivery apparatus. In reply to many disgruntled communities Civil Defence pointed out that they were not a government department as such and could only enable community safety and resilience but not provide it. This highlighted a fundamental dilemma or missing nexus in local emergency management: communities had to plan for and prepare to fend for themselves but Civil Defence, in spite of the putative but underdeveloped policy of well-resourced Sector Posts based in primary schools, could provide no bridging template or model for such self-sufficiency.

Community-Driven Response

From this unsatisfactory and unsettling state of affairs and following directly from the September 2011 meeting referred to above an independent non-statutory planning initiative was launched in Mt Pleasant in late 2011.[18] Discussions among an eight-member working party of local residents were held on a fortnightly basis between October 2011 and February 2012. It was decided at the first meeting in October that an emergency response plan for Mt Pleasant should be drafted in document form before the end of the year. With an academic background in urban and regional policy and planning and with a basic training in Controlled Incident Management Systems (CIMS) plus direct experience of rescue operations on the day of the February aftershock I offered to produce the first draft of such a document. This offer was accepted by the working party and I set about drafting the Mt Pleasant Community Response Plan.[19]

One of the first tasks I faced was to decide on the boundaries of the "response catchment." The local residents' association boundaries, the school zones, and the local government sub-ward boundaries all vary slightly. For statistics and data-gathering purposes and future comparisons I thought it best to try to align the primary or critical Mt Pleasant response area with the central government census "mesh block" of approximately 4,000 people to which a secondary care and response area located off the hill was added. In this way we could still account for the boundaries used by the other agencies and organisations. As a consequence a two-tiered boundary is set out in the current plan.

A community profile has also been drawn up in the plan using an informal strengths and weaknesses assessment. No particular preference is given to physical environment elements versus social characteristics and the emphasis is on straightforward, unsentimental description that avoids alarmism. The current strengths and weaknesses within the community are listed in Table 1. Although this level of detail will seem too specific to most outside readers it is important to remember that the information is intended for use by local residents rather than bureaucrats or officials in the first instance. Given its specificity the profile needs to be updated seasonally or at least annually as new information comes, e.g., because of further earthquake damage or major infrastructure repairs in the area. Furthermore, and in line with Civil Defence Group Plans, which do not go into great profiling details but which do include regional "hazardscape" analyses, the Mt Pleasant Response Plan includes its own local hazardscape list. This is set out in Appendix A. The information contained in Appendix A is useful for contextual and orienting purposes as it is intended to be read by locals and outsiders as a kind of primer for what risks one could expect to encounter within and adjoining the community.

How to Respond Before Official Help Arrives

However, from a response perspective the most vital part of the document is the section on who should do what in the event of an emergency. Not surprisingly, this assumes the rapid formation of a team of volunteers, the Mt Pleasant Community Response Team, to set up and run a Community Response Centre (CRC) based at the local primary school. However, it differs from official Civil Defence Group Plans or quasi-official Community Response Plans found elsewhere in New Zealand. First, the emphasis is on setting out the capabilities and functions required across six key areas rather than identifying particular individuals as occupying specific roles. Second, the principle of joint or dual leadership underscores each role in order to ensure a positive form of redundancy, i.e. it allows for turn-taking

Table 1. Strengths and Weaknesses Appraisal for Mt Pleasant.

Category	Strengths	Weaknesses
Topography/ Geographic setting	Runs from sea level to 500 m at Mt Pleasant itself. Most of the suburb sits between 0 and 300 m with a substantial part within the 25–150 m contours. Although the land rises steeply there are relatively few vertically sheer cliffs or sudden drops and the collection of housing on the hill could be likened to a large hillside Pa[a] with a flanking north-eastern valley section that descends to sea level	
Elevation	Generally good, not too high, with relatively little low-lying land vulnerable to flooding or storm surges	Areas above 200 m from sea level are prone to occasional snowfalls and icy roads which may restrict movements
Climate/Solar orientation	Very sheltered from prevailing winds in places Solar orientation is generally good as the suburb has north to west aspect which tracks the path of daytime sunshine	Southern slopes exposed to southerly winds and gales Eastern slopes exposed to stiff easterly Sea-level northern edge and particular ridge-lines exposed to prevailing cool easterly while southern flank on the edge of Heathcote Valley exposed to southerly gales and nor-west gale force winds will threaten north and west facing sections of the hill
Housing stock	Generally less than 60 years old, which provides coherence of a sort but (earthquake effects omitted for now) means a highly aged stock in 40 years with attendant deterioration and weathering vulnerabilities	
Infrastructure	Roading, stormwater, sewerage and electricity substations date mostly from after 1960s and were generally in good repair prior to earthquakes in 2011 Repairs since earthquakes appear to have stabilised the reliability of underground and overhead services	Roading pattern confusing and convoluted overall and main roads encourage downhill speeding and bad driving Several northern approach roads are completely blocked at lower levels due to earthquake damage, i.e. Maffeys Road, Quarry Road, Rangatira Terrace. CCC has given no repair timeframe or schedule for repair of

		these roads as at end of 2011. Mt Pleasant Road has at times been reduced to a single, temporary traffic light-controlled lane between Seamount Terrace and Billy's Track and appears vulnerable to further subsidence
Services (medical, school emergency)	Professionally staffed Medical Centre operates within the community during normal surgery hours	No dedicated emergency services, emergency supplies or police station (paid or volunteer), cf. Sumner and Lyttelton which have Fire and Police (and Ambulance in the case of Lyttelton)
	Primary/Intermediate school sits midway in the geographical unit (114 m elevation and at centre from aerial or plan view) – relatively safe site although exposed to easterly winds and nor-west which is potentially dangerous if the latter reaches gale force level. Has multiple modern, well-serviced rooms, seating, whiteboards and computers sufficient for command and control operations in the event of an emergency	Woolston Fire Station (Woolston is the nominal covering station for Mt Pleasant) and the Urban Search and Rescue base are currently located at the end of Olds Place off Charlesworth Street are both operating under compromised circumstances due to effects of liquefaction and damage to main station building
Open space/ Walkability/ Legibility as distinct suburb	Street names for most of the main roads accurately define the unit's identity as a distinct hill or mound, e.g. Mt Pleasant Road, St Andrews Hill Road, McCormacks Bay Road, Cannon Hill Crescent, and are not repeated elsewhere in the city which reduces the likelihood of confusion	The suburb can be walked over in its entirety, albeit with considerable effort as there are no obvious transit routes in parallel with the geographic contour lines
		No easy connection to upper hill natural assets – cut off from Mt Pleasant itself and walkways
	Three other useful open spaces at different elevations: McCormacks Bay Reserve (20 ha area; 2 km perimeter); King Park (1 ha area; 642 m perimeter); Old School Reserve (3 ha area; 752 m perimeter). These are potentially good assembly points depending upon threat and any need to sectorise hill population into emergency sectors or plan for overflows as a result of inflows from other suburbs, e.g. tsunami.	Parks probably do not function to their full potential (e. g. for social events on hill allowing people to familiarise themselves with assembly points)
		Mt Pleasant is currently a location not a destination when compared with Redcliffs, Sumner, Lyttelton, Diamond Harbour or even Governors Bay
		Very few level and dedicated staging/orientation/resting points for pedestrians or cyclists as they traverse the catchment

Table 1. *(Continued)*

Category	Strengths	Weaknesses
	Other open spaces Soleares Playground 452 sqm 295 mp; Major Hornbrook Reserve 2076 sqm 188 mp; Major Hornbrook Reserve No.2 4088 sqm 370 mp (verify location for this); Parnham Reserve 2126 sqm 209 mp.	
Landmarks/Focal points / Observation points	Good 180 degree surveillance points at higher elevations, e.g. Old School Reserve, from where assessments of incoming threats and observations useful to other parts of city can be mad, e.g. advance of fire front	No shopping centre as such (but Mt Pleasant Medical Centre, Pharmacy, Hair Salon and Restaurant at bottom of Soleares Avenue). Other than school and temporary community centre no buildings practical use for evacuation points, supply depots, command post or other emergency purposes No unimpeded view of a natural topographical or geographical focal point within the suburb which makes for a logical community hub No internal natural or human-made landmarks visible to all
Demography	High income suburb (rated Decile 10 on 1–10 socio-economic scale) with relatively low crime rate Low ethnic diversity and no documented ethnic tensions	Older population (45 years of age on average) compared to rest of city – aged resident population and working relatively new arrival families without ties beyond school; many professional couples without children not home that often
Accessibility	Potential to for people cross over into Lyttelton Harbour on foot and avoid use of roads, e.g. invasion by hostile forces, pandemic	Main road on northern boundary acts mainly as a commuter/sightseer vehicle conveyer belt which breaks the connection with a major natural asset (the estuary) and will become a traffic bottleneck in emergency conditions if panic evacuations occur

[a]"Pa" is a traditional Maori term for a fortified or defended village.

and down-time between leaders, and it assumes that conferring between two parties will yield better decisions than unilateral calls under pressure by a single individual.

The key roles to be filled are as follows: Joint Overall Supervisors; Joint Logistics Co-ordinators; Joint Welfare and Resource Co-ordinators; Joint Neighbourhood Support Co-ordinators; Joint Medical Assessment Co-ordinators; and Joint Communications and Liaison Co-ordinators.[20] The types of roles and tasks vary from role to role. As an example the Joint Logistics Co-ordinators' tasks are outlined in Appendix B. As can be seen from the list in Appendix B the role demands at first glance may seem daunting. Indeed, the number and type of tasks is similarly high across all of the portfolios.

Pragmatism in a Pluralistic Society

Importantly, however, the distinction between this plan and those used by Civil Defence and other agencies is that these roles do not require accreditation or licensing or formal training. To use a metaphor they are "shoes to be filled" in times of need. The underlying assumption is that "civic expertise," as Flyvbjerg (2001) and others have discussed the concept, already resides in the community and in everyday life (Schafer, Carroll, Haynes, & Abrams, 2008). Expertise is not just a function of professional and scientific training. Instead, and more in keeping with the philosophy of guerrilla warfare and underground movements rather than conventional military thinking, which seems the shadow the concept of Civil *Defence* (my emphasis) in many countries, the assumption is that an event will trigger the appearance of competent and often anonymous individuals to do the work required.

The main principle, then, is one of trust in the existing social capital and civic expertise to be found within the local community rather than identifying professional or quasi-professional skills that need to be gained, with the exception of medical training which must always be accredited in some way, through institutional learning, i.e. Civil Defence courses. There simply are not enough people willing and able to act as volunteers as Civil Defence would prefer and this malaise of declining commitment to voluntary organisations and their cultures is not confined to this particular entity but is a common trend in an increasingly pluralist society.

In the case of the Mt Pleasant Community Response Centre it is expected that a core group of five to six people, most likely, but not necessarily, drawn from the original working party, will get the response centre up and running. The local primary school is the location of first choice, although alternatives are specified in the Plan, and a Memorandum of Understanding exists between the school Board of Trustees (along with the School Principal) and the local residents' association that allows the use of the school in emergency circumstances.[21]

Standing By for the Next Event

Since February 2012, the point at which the Plan was launched, there have been no incidents that have necessitated a full or even partial activation of the response centre. There are multiple keyholders for the school and a roster has been used at lengthy holiday times to ensure access. Although Civil Defence has been provided with a copy of the Community Response Plan and senior staff have been able to provide feedback on its content it does not have any statutory standing, nor is it officially endorsed by Civil Defence.

This is partly due to the fact that since February 2011 Civil Defence has itself been subject to a number of internal and external reviews and restructuring processes many of which have yet to be finalised. At the same time the organisation is still considering how best to engage with local communities and provide appropriate support. It seems fair to say that the present context is one of renewed efforts in trust and capacity-building but there is a degree of wariness and weariness across the spectrum of emergency management planning institutions. For the present time it seems a mutually acceptable compromise for the local community to have an informal but still structured plan in reserve.

RISKS TO THESE GAP-FILLING INITIATIVES

Constant Flux

The first major risk, as mentioned earlier, is the general "brew" of volatility around employment, housing, insurance payment delays, home repair delays, antagonisms between central and local government, the findings of royal commissions, ambitious and sometime conflicting new master plans for the city, perceived lack of transparency in government and corporate decision-making processes, looming school closures due to population decline, "sudden" collapses in supposedly solid businesses and institutions, individual mental and physical health issues, summary closures of newly discovered earthquake prone building and general fatigue and exhaustion. For example, in the case of GtR changes in employment status and living arrangements for both the key project co-ordinator and for the Chair of the GtR Trust will mean the departure of these "veteran" members as of mid-2013. The Mt Pleasant initiative has been affected by the demolition of the existing Community Centre that belonged to the Residents' Association insofar as the rebuilding plans have caused tension with the local primary school which also needs to rebuild its damaged school hall and expand its site. At times it appears to be a battle as to who will own the new community hall rather than what is best for the community as a whole. Predictably, membership of the Mt Pleasant Community Response Plan team includes key figures from the Residents' Association and the School Board of Trustees so relationships can be frayed at times.

The background environment of constantly changing circumstances poses more specific threats. The closure of Mt Pleasant School, e.g., might be extremely problematic for community response planning, especially in the absence of a community centre and local hall, but it would not jeopardise the activities of GtR. Nor does the proliferation of earthquake recovery groups who will have to compete for limited funds threaten the ability of Mt Pleasant residents to set up a community defence facility. But while GtR is unlikely to be made redundant by big-scale changes in emergency management around natural disasters it could be overshadowed by big projects that seek or promise to green the damaged city. The proposed "green frame" produced recently by the CCDU or the Avon Otakaro River red zone park may be enough to make greened corner store rubble sites look rather tokenistic and ramshackle. The Mt Pleasant initiative might suddenly find itself undermined by larger institutional changes or rebranding as almost all of the emergency response bodies in New Zealand have gone into review mode or have been brought to account before a royal commission or an inquest.

New Versus Old Institutions

Second, the net effect of the coronial inquests and Royal Commission into human deaths and building collapses is that many authorities have been driven into a frenzy of health and safety paranoia. This makes it difficult to get approval and support for many new initiatives in case new or unanticipated liabilities might arise. Third, there are the political and ethical dimensions of seemingly apolitical new institutional arrangements. For example, many see the greening or better interim use of vacant city sites as a matter of policy, regulatory and spending priorities within municipalities. GtR could be seen as letting the city council off the green hook, so to speak. Also if GtR has no ecological principles or planting manifesto and simply aims to get vegetation into vacant sites as quickly as possible all sorts of inappropriate and potentially dangerous plants could be introduced.

For the Mt Pleasant group the implicit political critique of organising something locally because there is a perceived gap in community protection at local government and central government level has its own perils. Antagonistic, elitist and inward-looking stances might evolve within the group. There is a risk, albeit a highly remote one in urban New Zealand, that a community of survivalists who then start trying to indoctrinate or weed out "weaker" members of the community will actually cause more harm than good. High social capital is not intrinsically worthy simply because it exists.

Civics and Citizenship

In a related vein there is the question of whether civilian protection from disasters should be left to volunteerism and inadequate funding from central and local governments when the same putative organs of the public good seem content to direct local rating and general tax money to the underwriting of major professional sporting and business events and facilities. This raises a more fundamental question about the changing role and identity of the volunteer in New Zealand society. As noted earlier, although the volunteer-based clubs and associations of the past may have started to wane in popularity the willingness to help seems to run deep. There is always the risk of stretching that fabric too far and jeopardising worthy but hard-to-market causes by drawing potential volunteers into trendy and transient projects. Even assuming that a large reserve of volunteers exists, the questions arise around the sorts of volunteers that are needed and how to attract them in useful proportions. It may be particularly the case in Christchurch at present due to general "quake fatigue" but anecdotally it seems hard to find people with both long-term vision and day-to-day problem-solving capabilities at present.

The emergency management agencies see a different issue with respect to volunteers: they simply do not seem to know what to do with them, hence the Student Volunteer Army and Farmy Army phenomena. Across a range of documents and communications the agencies have labelled these individuals "spontaneous volunteers," which has connotations of the "instant mob" rather than skilled or capable people making themselves available by using sound intuition. Institutionalisation and enrolment through formal training (and most probably very careful psychometric testing) is the only response that these agencies can produce to date. By contrast both GtR and the Mt Pleasant Community Response group regard volunteers as people who can be quickly mobilised to offer skills and expertise if they are shown a basic level of trust and are given appropriate cues for activation. The distinction here is between community faith in mobilisable skilled individuals and institutional paranoia about possible untrained or partly trained amateurs.

The Urge to Forget

There are other more specific factors. Reduced threat perception is starting to return in local communities. The Mt Pleasant Community Response team needs to move beyond paper-based planning only and run some exercises but it seems unfair to put people through an emergency exercise when what they prefer most is calm and stability in the neighbourhood. The intended exercise for late 2013 is a tsunami refugee scenario with emergency response agencies observing and supporting.

For GtR reducing tolerance over time for the "temporary chic" aesthetic, where improvisation is appreciated even if things look a little too informal by conventional standards, may become a significant issue. Already, in Christchurch the novelty value of shops and offices being located in shipping containers has started to wear thin as people complain of the monotony of narrow rectangular boxes in which they must work or shop. Temporary gardens may be interpreted as mean-spiritedness on the part of civic bodies or may interfere with the plans of groups who seek permanent native ecological restoration sites in the city. Similarly, English or European garden and park advocates may see GtR as an impediment to the re-establishment of Christchurch as a classical or picturesque garden city.

When to Quit

There is also the matter of institutional inertia that any crisis-generated organisation will face over time. GtR is a socially constructed entity born unintentionally out of physical events. In this sense it matches Hacking's view of social constructions as things created in order to eliminate undesired or untoward circumstances (Hacking, 2000). In theory GtR should fold when the rubble has been cleared and the city has been rebuilt but there is no exit strategy as yet. If the Mt Pleasant model is seen to work and becomes adopted elsewhere it might then be taken up as more fully funded programmes by government bodies. This development, together with more successful individual household preparedness campaigns run by Civil Defence or the Mt Pleasant group, would render the response plan redundant. Again, an exit strategy has not been prepared. In actual fact both groups, if they survive other challenges, are likely to mutate into resilience-building or community-building entities where broader urban ecology concerns would occupy one while community maintenance and enhancement would become the focus for the other including such issues as how to make the suburb more of an adventure space for adults and children alike.

CONCLUSION

Discussion around resilience has in recent years evolved from earlier and more narrowly framed mechanistic and ecological theories to continue to account for human intentionality and emotions. For example, Frommer (2011, p. 96) argues that resilience "implies three qualities, these being 'resistance', 'recovery' and 'creativity'" (Frommer, 2011). The more operationally focussed disaster management literature has produced at least one four-phase conceptual model or sequence. Widely employed by emergency management organisations, including those in New Zealand, it is commonly referred to as the "four R's" model: reduction (also referred to as mitigation in some formulations); readiness (sometimes labelled preparedness), response (and relief); and recovery (rehabilitation).

In this context, and in light of the "local responses" theme of this volume, GtR can be best characterised as a creative, and to some extent "playful," recovery activity (with a longer ecological recovery or restoration trajectory nested within its immediate urban rehabilitation ethos) while the Mt Pleasant Community Response Plan straddles an area composed of resistance, readiness and response. The latter initiative looks ahead to gauge threats and vulnerabilities from year to year while providing a backup plan if those threats materialise with major impacts that are not, or cannot be, met immediately by official emergency response procedures from government institutions.

At the same time there are questions and challenges around democracy and participation. The custodians of the new grass-roots initiatives discussed above do not want to cut across or ignore the principles of participatory democracy as advocated, e.g., by the International Association of Participation in its "IAP2 spectrum of participation": inform; consult; involve; collaborate; and empower.[22] There is the constant risk that a new grass-roots project simply becomes a therapeutic exercise for those leading the initiative. Conversely, leadership, risk-taking and the launching of provocative ideas and projects are critical to success in new social movements.

There is also the matter of correctly gauging the social capital and resilience thresholds within communities and not assuming or expecting too much. The notion of resilience, although widely used, is far from satisfactorily defined. Lorenz (2010, pp. 7–12) argues for a tripartite definition: adaptive capacity; coping capacity; and participatory capacity. The first two capacities are largely reactive in this interpretation (Lorenz, 2010). Read this way the situation in Christchurch almost three years after the earthquakes would locate most people in the first two categories with a concomitant high degree of overall vulnerability. This should signal significant caution in trying to launch and maintain any potentially long-lived projects. Conversely, Franklin (2008, pp. 32–34) notes that the deadly bushfires on Black Saturday in Hobart in 1967 helped to galvanise and ensure adequate resourcing of volunteer rural fire brigades in Tasmania (Franklin, 2008). The Tasmanian response still serves as a model for the rest of Australia and furnishes proof that not all catastrophic events drive people primarily into survival or coping mode for extended periods of time.

Initiatives like the Mt Pleasant Community Response Plan should avoid perpetuating what Gunder (2008) describes as the "hauntology of planning" in relation to risk. Managing for risks as if they are matters only of objective science and technical procedures without acknowledging the socially constructed and human emotional dimensions of risk perception will not in the long run make a great difference (Gunder, 2008). For better or worse what an event and its aftermath such as the Christchurch earthquake allows is the observation of the full panoply of responses across individuals, collectives and institutions and some measures can only be designed as these responses play out. In other words, resilience is always fraught as Hutter, Kuhlicke, Glade, and Felgentreff (2011) have observed (Hutter et al., 2011).

As a closing remark and modest contribution to understanding what happens during natural disasters, and this is based on my own experiential reality rather than an academic appetite or need for a comprehensive or credible response typology, I suggest that local responses to catastrophic untoward events need to be able to anticipate and accommodate a flux of nine individually and collectively experienced factors which I will call the "nine f's of response": fight (ready to defend/adopt defensive stance), flight (ready to flee/escape), fright (traumatised into complete inaction); fragmentation (narrow-focus capability only), floating (vague alacrity), fazing (irritating but not entirely debilitating confusion); fronesis (ability to quickly apply practical wisdom), fusion (unwavering collaborative attitude), futuring (unwavering forward-looking attitude). The challenge, both for GtR and the Mt Pleasant Community Response team, is to learn to live with this reality for the foreseeable future.

NOTES

1. http://www.stuff.co.nz/national/christchurch-earthquake/4819335/Second-Christchurch-fault-much-better-behaved

2. The current figure for "red zoned" properties, i.e. condemned house and land, is 7,800. http://ecan.govt.nz/publications/General/LURP-Context.pdf. p. 11.

3. More indirectly I have been involved in on-going emergency "local response" activity for the past nine years by virtue of being a volunteer firefighter, and local volunteer firefighters were very much at the forefront of the immediate natural disaster response in September 2010 and February 2011.

4. For the sake of a broad comparison I have used a European country and because this chapter is based on a conference presentation at the International Geographers Congress held in Cologne in 2012 it seemed logical to select Germany for that purpose.

5. See http://www.nfpa.org/assets/files//pdf/os.fdprofile.pdf

6. See http://www.sva.org.nz/; https://www.facebook.com/pages/Farmy-Army-Federated-Farmers/203178676360140; http://www.cancern.org.nz/

7. The term "Garden City," therefore, when used in relation to Christchurch errs very strongly on the side of visible "order," making even the "messy ecosystems, orderly frames" and "cues to care" strategy of hiding wildness behind picket fences espoused by Nassauer, for example, seem a bridge too far for most citizens (Nassauer, 1995; Greenep, Montgomery, Sullivan, & Meurk, 2010).

8. I joined the Steering Group at this meeting.

9. See http://greeningtherubble.org.nz/wp/?p=224 and http://greeningtherubble.org.nz/wp/?p=128 respectively.

10. See http://greeningtherubble.org.nz/wp/

11. For example a culture and arts group called Gap Filler began in late 2010 and a site-use brokering agent called Life in Vacant Spaces was established in 2012.

12. The first project, Victoria Green, was a relatively high specification exercise despite its simple appearance. Approximately $20,000 in materials and services were donated.

13. It became apparent over the first year of operation that participants have often been too busy on particular sites to provide durable on-site signage and helpful interpretation to the extent that some sites had no clue at all as to who had created them.

14. http://www.stats.govt.nz/Census/2006CensusHomePage/QuickStats/AboutAPlace/SnapShot.aspx?pdf=1&id=3596000&type=au&ParentID=1000013&expand=2000060&scrollLeft=0&scrollTop=261&ss=y

15. A "portaloo" is a portable toilet cabin usually transported to places such as building sites or outdoor concerts where there are limited permanent sewage connections.

16. The reticulated city water supply, including reservoirs and hundreds of kilometres of pipework, was severely damaged and not reconnected for nearly three weeks.

17. People also dropped off camping supplies. At the time of writing the school has a shed full of unclaimed equipment, some of it very high-quality camping gear, which sits in reserve for future emergencies.

18. I was contacted at this point by a member of the Mt Pleasant Residents Association on the basis of my experience as a volunteer firefighter.

19. The initial working title was the Mt Pleasant Emergency Response Plan but Civil Defence did not like the se of "emergency" so it was changed to Mt Pleasant Response and Resilience Plan. The decision to use "community response plan" came from the realisation that there were other such plans in circulation in local communities in New Zealand but more importantly that it was too ambitious to combine long-term resilience planning with short-term disaster event planning in a single document. It is important to note that the Community Response Plans created to date locations throughout New Zealand range from a post-quake list of eight actions done to elaborate coastal evacuation plans written under the close supervision and oversight of Civil Defence staff.

20. These roles and tasks are not dissimilar to those used by Civil Defence in the running of larger Emergency Operations Centres or Area Co-ordination Centres.

21. Strictly speaking the Ministry of Education owns the land and buildings of all state schools in New Zealand but School Boards of Trustees have authority for day-to-day operational and budgeting functions.

22. http://www.iap2.org/associations/4748/files/spectrum.pdf

REFERENCES

Flyvbjerg, B. (2001). *Making social science matter: Why social inquiry fails and how it can succeed again.* Cambridge, MA: Cambridge University Press.

Franklin, A. (2008). A choreography of fire: A posthumanist account of Australians and Eucalypts. In A. Pickering & K. Guzik (Eds.), *The mangle in practice: Science, society and becoming* (pp. 17–45). Durham: Duke University Press.

Frommer, B. (2011). Climate change and the resilient society: Utopia or realistic option for German regions? *Natural Hazards*, *58*, 85–101. doi: 10.1007/s11069-010-9644-0

Greenep, H., Montgomery, R., Sullivan, J., & Meurk, C. (2010). Rethinking stony ground: Urban ecology, native plants and 'greyfields.' In D. Nichols, A. Hurlimann, C. Mouat, & S. S. Pascoe (Eds.), *Green fields, brown fields, new Fields: Proceedings of the 10th Australasian urban history, planning history conference, University of Melbourne, 7–10 February, 2010*. Melbourne: University of Melbourne/Melbourne School of Design (pp. 159–172).

Gunder, M. (2008). Ideologies of certainty in a risky reality: Beyond the hauntology of planning. *Planning Theory*, *7*(2), 186–206. doi: 10.1177/1473095208090434

Hacking, I. (2000). *The social construction of what?* Cambridge, MA: Harvard University Press.

Hutter, G., Kuhlicke, C., Glade, T., & Felgentreff, C. (2011). Natural hazards and resilience: Exploring institutional and organizational dimensions of social resilience. *Natural Hazards*, doi: 10.1007/s11069-011-9901-x

Lorenz, D. F. (2010). The diversity of resilience: contributions from a social science perspective. *Natural Hazards*, doi: 10.1007/s11069-010-9654-y

Montgomery, R. (2012). Greening the rubble in Christchurch: Civic ecological reclamation efforts during a crisis event. *Lincoln University Planning Review*, *3*(2), 3–13.

Nassauer, J. (1995). Messy ecosystems, orderly frames. *Landscape Ecology*, *14*(2, Fall), 161–169.

Putnam, R. (2000). *Bowling alone: The collapse and revival of American community*. New York, NY: Simon & Schuster.

Schafer, W., Carroll, J., Haynes, S., & Abrams, S. (2008). Emergency management planning as community work. *Journal of Homeland Security and Emergency Management*, *5*(1), Article 10.

APPENDIX A: MT PLEASANT HAZARDSCAPE

- Earthquake (local) – expected aftershock period in existence.
- Earthquake (alpine fault).
- Heavy rain-induced landslides or rockfalls/erosion of the Port Hills – earthquake slips increase slope hazard areas.
- Vegetation fires – on rural margins or in reserves where routine maintenance and grazing have been interrupted or unmaintained and abandoned urban properties.
- Water supply failure – fragile and intermittently out of action since earthquakes.
- Wastewater failure – fragile and intermittently out of action since earthquakes.
- Electricity failure – fragile and intermittently out of action since earthquakes.
- Tsunami – indirect for most residents but direct for some residents.
- Snow/ice – August 2011 severe snow event demonstrated vulnerability.
- High winds – including loose materials from already damaged homes.
- Urban fires – within abandoned or damaged structures.
- Workplace accidents – as dozens of private properties become construction sites during repair and rebuild from earthquake chances increase for sparks igniting structures; vegetation; heavy equipment accidents; nail-gun and saw injuries.
- Major road accident – as more heavy vehicles are used on a restricted number of roads; as people mistakenly try to use blocked roads; bottlenecks in roads that are open, i.e. Glenstrae Road, Major Hornbrook Road, Mt Pleasant Road and Cannon Hill Crescent.
- Regional drought – local effects heightened if water restrictions remain in place for extended periods and throughout summer.
- Human disease pandemic – immune systems of residents compromised due to stresses from earthquakes, e.g. respiratory conditions due to dust.
- Civil unrest – trauma caused by land zoning decisions plus deteriorating global economic conditions leads to mob events and more opportunistic crime in depopulated suburbs.

APPENDIX B: JOINT LOGISTICS CO-ORDINATORS' TASKS

These are as follows:

- They will deal CRC logistics and forward planning and will make up jobs to be done out in community and gather lists of supplies and services needed for both the CRC and in the community, e.g. "portaloo," i.e. portable toilet, needs.
- They will normally pass all requests for materials and equipment through the Joint Communications and Liaison Co-ordinators.
- They will routinely work with volunteers who have identified themselves as professional, qualified or highly experienced tradespeople, contractors, drivers and otherwise practically skilled individuals to carry out particular jobs involving movement of supplies and people, the fabrication of temporary structures, setting up of generators and other general labouring activity should it be required.
- They will work with public utilities (electricity, water supply, wastewater supply) and infrastructure (roading) personnel and assist them or direct them as required.
- They will prepare the daily and/or weekly rosters for CRC staffing during an emergency.
- They will prepare material for the daily briefings by way of written and/or verbal situation reports.
- They will manage on-site security including the delegation of gate staff and building safety and security and vehicle traffic flow in and around, i.e. on the street, the CRC.

CHAPTER 4

THE ROLE OF SOCIAL CAPITAL IN POST-FLOOD RESPONSE AND RECOVERY AMONG DOWNSTREAM COMMUNITIES OF THE BA RIVER, WESTERN VITI LEVU, FIJI ISLANDS

Othniel Yila, Eberhard Weber and Andreas Neef

ABSTRACT

Floods are among the most significant and frequent hazards to affect communities in the downstream part of the Ba River in Western Viti Levu, Fiji Islands. They often leave in their wake displacements and death putting thousands at risk of sliding into poverty. Using the recent 2009 and 2012 floods, we examine how social capital aids in post-disaster response and recovery among residents in five selected villages in the downstream communities of the Ba River. Data were collected from a questionnaire survey administered to 97 households and semi-structured interviews with a further 20 respondents. It is conventionally

Risks and Conflicts: Local Responses to Natural Disasters
Community, Environment and Disaster Risk Management, Volume 14, 79–107

ISSN: 2040-7262/doi:10.1108/S2040-7262(2013)0000014010

believed that moving supplies, aid and expertise into flood-affected areas offers the best path to effective response and recovery. By contrast, our results indicate that residents of downstream communities in Ba District are using four approaches to create and deploy social capital among them to facilitate disaster response. The patterns of social capital used for effective response include practices of search and rescue, information, mutual assistance and commercial cooperation. Such strategies help to build resilience at household and community levels and reduce risks of loss of life and costly damage to property. The findings can be used to generate policies concerning the integration of social capital as a component of flood disaster response and recovery mechanisms.

Keywords: Disaster recovery; flood; social capital; Ba River; Fiji Islands

INTRODUCTION

Flood disasters are not only physical hazards, but also social ones. They help us to better understand complex social systems. Natural hazards, such as floods, give clues where a society is vulnerable and where it is buoyant. It also helps to better understand the degree of social cohesion in efforts towards recovery among the affected communities (Appadurai, 1984). Among all types of disasters, floods have annually caused the greatest loss of life and property. They have threatened the strength and stability of communities around the world (Adler, 2009; Zhao, 2010).

Most media coverage on flood disasters focuses on the extent of physical damages they inflict: loss of lives, damaged infrastructures, ruined subsistence and commercial agriculture and deleterious effects on public health. Many hold the view that the extent of damage determines the speed of recovery (e.g. Aldrich, 2010, 2011). A common assumption is also that the larger the scale of government response, and the more assistance and money is flowing into affected areas, the more viable will be the path towards successful recovery for affected populations. Hence, the response to flooding by government agencies, non-governmental organizations (NGOs), faith-based organizations (FBOs) and relief agencies has often been premised on the perspective that deploying more financial and material supplies, personnel, and experts into the affected areas will enhance the state of recovery (Aldrich, 2010, 2011).

The Fiji Islands are no strangers to the caprices of devastating flood disasters. Yeo, Blong, and McAneney (2007) chronicled a 100-year historical series (1892–2002) of flooding in Fiji. His records attest to the popularly held view that the frequency of flooding has increased in recent decades causing disruption of the normal functioning of society and its sub-units with attending amplification of physical, economic and social risks (Yeo et al., 2007; cf. Duaibe, 2008; Mudgway, 1997; SOPAC, 2009; Terry, McGree, & Raj, 2004). A recent report by the United Nations Office for Disaster Risk Reduction (UNISDR) and the United Nations Development Program (UNDP) shows that the frequency of high-intensity floods is on the increase in Western Viti Levu with devastating floods which used to occur every 190 years, now projected to occur every 25 years by the year 2100 (or even more often). In contrast to these findings and predictions, however, stands the fact that so far the 1931 flood in Ba and Nadi has been the worst in terms of flood levels and casualties. Even the floods of March 2012 did not reach the levels of 1931. One could conclude that floods in the western part of Viti Levu have increased in frequency, but not necessarily in intensity.

Flood disasters will continue to happen and damages and losses accompanying them cannot be completely eliminated even if communities are better prepared and warned earlier of approaching flood hazards. In view of governments' increasingly overstretched budgets and the outlook that flood disasters are likely to increase in frequency in Fiji, this chapter highlights the critical role of social capital as one component of effective community response to withstand such hazards and avoid or minimize disasters. Examining post-disaster responses of residents in Ba District, the chapter argues that those with robust social networks are in a better position to coordinate recovery whenever floods hit them.

The contribution then seeks to make two principal points. First, it adds to the small but growing literature on social dimensions of floods in the Pacific by analysing the under-explored intersection between flood disaster and social capital. Most publications on floods in the Pacific have been focusing on technical aspects of flood prevention and management (SOPAC, 2002, 2009; Yeo, 2000). We argue that adding a social science perspective on flood disasters can provide better tools for improved management policies. Second, the chapter draws an important link that social capital available to victims of floods matters most immediately after the flood event (emergency response phase) and in the long period of post-disaster recovery (restoration phase). When floods strike the first respondents are not trained emergency workers, but local residents. They have deep knowledge of the area and they arrive well ahead of media, government, NGO, and relief agencies. In the days, weeks, months and years of recovery social capital and networks continue to be a critical resource, serving as "social insurance" and providing victims with information, financial help and physical assistance. Social capital employed by individuals, households and the community will provide insight to new ways of promoting preparedness as well as recovery. Flood disaster recovery is not only about rebuilding destroyed houses and infrastructure, but also about the reconstruction of whole communities building safer places. Social capital remains crucial to help mobilize members of communities towards collective action for recovery. Social capital has an important role to play in creating and maintaining resilience at the grassroots level.

While there has been a growing recognition of the value and place of social capital networks in our everyday lives, the incorporation of social capital programmes and policies in flood disaster recovery management has been rare and often overlooked by governmental and NGOs. There is little evidence that decision makers and NGOs take into consideration the importance of social capital during flood recovery. Discussions conducted with representatives of government agencies and NGOs revealed that the talk of the day is community resilience, community-based adaptation to natural hazards, and the strengthening of social cohesion and social capital. On the other side detailed ideas how to achieve this and what to do differently from the way disasters have been addressed so far remain rather scant and general. Innovative ideas are rare. Our research intends to address this gap starting by gathering information how social capital already now is "at work" when hazards strike. We then conclude and recommend that it is important to put social capital on the agenda of decision makers, and relief and rehabilitation managers. It is hoped that the insights revealed through our research help to initiate discussions and create future plans that are more effective than the present ones in contributing to disaster recovery. This chapter therefore seeks to demonstrate that social capital among residents of flood-affected communities in Ba District remains the most crucial component in the flood-disaster recovery process. The quality of community cooperation creates cohesiveness and resilience that is more stable and sustainable than the often short-lived involvement of outsiders.

THEORETICAL FRAMEWORK: SOCIAL CAPITAL IN DISASTER RECOVERY

Social capital is one of those concepts that seem to defy consensus usage and applications. However, at its core, it provides a better explanation of how people respond to natural hazards such as flood through the use of their relationships to other actors in societies for their own and for the collective good. To buttress this position, Hawkins and Maurer (2010, p. 1778) posit that "social capital has gained intellectual currency as a means to understand the relative strength of families and communities in time of disaster."

Although the formulation of a conceptual framework of social capital from an academic perspective is often attributed to Pierre Bourdieu (1986), the groundbreaking research of Robert Putnam (1993) helped to popularize the concept. Putnam defined social capital "as the feature of social organization, such as trust, norms and networks that can improve the efficiency of society by facilitating coordinated actions" (Putnam, 1993, p. 27). Coleman's (1988) position of social capital as cooperative relationship within families and communities and Putnam's perspective on civic engagement, trust and norms of reciprocity have remained the dominant contributions to the social capital discourse for years. In our view, Coleman's conceptualization of social capital has a higher utility for social workers, as his explanations portray social capital as "utilization of direct and indirect resources that arises as an outcome of social network and social support system among family, relatives, friends or community" (Coleman, 1990; cf. Hawkins & Maurer, 2010, p. 1778).

Lately the role of social capital has been further explored by many scholars in three major disaster recovery processes: the Kobe Earthquake of 1995, Hurricane Katrina and its fury particularly in New Orleans in 2005 and the Indian Ocean Tsunami in December 2004 (Aldrich, 2010, 2011; Boettke et al., 2007; Chamlee-Wright, 2010). The findings revealed that New Orleans did not witness the vibrant recovery process that happened in post-tsunami South and Southeast Asia in spite of its clear material advantage. Many attribute the much slower recovery to a social capital deficit (e. g. Aldrich, 2010).

Researchers have differentiated three forms of social capital based on quality and the way they work and become effective: bonding, bridging and linking social capital. The bonding social capital strengthens relationship that exists among like-minded people or members of a network connected together through similar activities or organizations (Aldrich, 2011; Hawkins & Maurer, 2010), people that have few social differences, people that are connected to each other horizontally, endowed with similar resource sets and holding similar positions and status in society. It comprises strong horizontal ties which connect members by providing a safety net. Putnam (2000, p. 22) regards such form of social capital as "undergirding specific reciprocity and mobilizing solidarity" or what Szreter and Woolcock (2004, p. 650) attribute as "reinforcement of homogeneity." Bonding social capital entails strong ties that exist among relatively small groups of people. Bridging social capital refers to the horizontal ties which connect people from different networks and groups and generates broader identities and linkages to external assets. Bridging has a host of other benefits for individuals, communities, societies and governments and often gives room for improving livelihoods and upward mobility. It builds the basis for people who are dissimilar in social structure and who may have little if anything in common with one another to cooperate and share resources with an endearing social change (Sopha, Oeur, & McAndrew, 2007). The third type, linking social capital, refers to the vertical ties that individuals builds to connect them with other individuals and institutions that have relative influence and power over them such as government agencies, banks and elected officials as to grant them accessibility to services, resources or jobs (Hawkins & Maurer, 2010; Szreter & Woolcock, 2004). Summarizing the forms, Aldrich (2011, p. 4) holds that "where bridging social capital connects individuals of more-or-less equal social status, linking social capital connects those of unequal status, providing them with access to power. Linking social capital brings together citizens with decision makers and leaders who hold positions of authority and can distribute often scarce resources." People who possess strong networks within and across all three types of social capital are generally considered to be better able to cope with hazards such as flood.

Further debate among scholars has been whether social capital is a community or an individual asset (Hawkins & Maurer, 2010; Putnam, 2000) or a function of both (Coleman, 1990; Szreter & Woolcock, 2004). In our position social capital is based on actions by individuals in relation to structural forces and processes of society. Within this context the social capital framework receives meaning; it is the outcome of agency, of structural properties of society under which individuals act. Although we concur with the Hawkins and Maurer (2010) about the uniqueness of social capital to bridge existing theoretical gaps that may arise about the understanding and importance of individual action and community structure from either micro or macro angles, the quality of social capital lies within individuals, but it forms (and alters) society or – in our case – communities.

Another contention in literature has been whether social capital is a wholly positive asset, as not all forms of social capital guarantee a positive outcome of a desired type (Putnam, 2000; Coleman, 1988). On the one hand, studies by Aldrich (2011), Schuller, Baron, and Field (2000) and Hawkins and Maurer (2010) established that social capital emanating from support through social network can be beneficial to withstand personal, economic and emotional challenges. Others found disadvantages and downsides to social capital asserting that the very characteristics of social capital that enable productive benefits also have the tendency to bring negative externalities. Maintaining social networks can be time-consuming and costly and the benefit is not always clearly visible. Membership in social networks can act as a barrier to social mobility and inclusion, leading to division rather than bringing communities or societies together (Hawkins & Maurer, 2010; Schuller et al., 2000).

Although the importance of social capital has been emphasized in the functioning of Fijian society due to the resource tenure structure and communal nature of living, much of the work done has centred on the role of social capital with regards to community-based resource management with no work pointing to the role of social capital in disaster recovery (Sami & Reddy, 2011; Sano, 2008; Veitayaki, Aalbersberg, Tawake, Rupeni, & Tabunakawai, 2003). In his work Sano (2008, p. 3) states that "kinship groups and village or district-level activities make up the heart of social interactions, but other forms of association stretch beyond blood ties to include all *Kai Viti*, or Fijian people." Similar to tribes in Papua New Guinea, he showed that "when two Fijians meet for the first time, they will follow a standard pattern of small talk to figure out how the two are linked" (Sano, 2008, p. 5). Even if they find no direct kinship bonds, there are other socially important connections that they can rely on based on geography and history. For instance, if a man from Nadroga/Navosa province and the island of Vanua Levu meet, they will call each other *Dreu*, signifying a historical traditional link between the two regions, often following lineage back to ancient origin gods, or *Kalou Vu*. He showed that "if the Nadroga man meets someone from Kadavu Island, they will regard each other as *Tau*, a reciprocal relationship that brings about teasing and a sense of closeness greater than the *Dreu* relationship. To be respectful, it is important for a Fijian to recognize these links and accept the stranger as kin. In this way, all Fijians are connected to all other Fijians through historic bonds that continue to be important today" (Sano, 2008, p. 6).

Given that social capital is not uniformly manifested across time and societies, there is need to put into consideration the historical and cultural environment under consideration. With regard to the cultural diversity and heterogeneity of Indian society and institutions, Serra (2001, p. 699) positioned that "standard Western measures of social capital such as those used by Putnam in his 1993 study of northern and southern Italy – including literacy, voter turnout, and membership in horizontal associations – did not map well onto the empirical realities of Indian states."

In conclusion, social capital in whatever dimension and shades can be seen as a resource, an asset. Its utilization can enable people to achieve goals they would otherwise be unable to attain. It is undeniable that people often fall back on their network of social capital to obtain access to information and resources. On the other, they utilize their knowledge governing the behaviour and norms within a particular network to give them perspectives of how other members are likely to respond as well as to guide their decision on how they should act. Thus, social capital enables people to effectively exploit available opportunities at their disposal, both within "free markets" and also in "non-market settings" (Fukuyama, 1995; Sopha et al., 2007). In addition, by drawing on social capital, people have the potential to reproduce and at occasions transform the network of social relations context they operate (Chamlee-Wright, 2010). Finally social capital can link people to power centres, to centres of decision making, and thus allow them to – albeit indirectly – participate in power and contribute to the decision making process. All these properties are crucial particularly when external shocks create havoc to people's lives, when people have been materially and emotionally harmed and wounded, and when public response systems are under severe pressure and stress, stretched to their limits what manpower and resources are concerned. Especially in such situations it is crucial to have alternative support structures. This is the time when social capital is to prove its worth.

METHODOLOGY

The main fieldwork for the study was conducted in five villages of Ba District of Western Viti Levu in February 2012, immediately after the January 2012 flood. The five villages were selected after consultation with the Disaster Management Office at the Provincial Office. They were the most affected villages during the 2009 and 2012 floods. A follow-up fieldwork was conducted in one of the five villages (Votua) in September 2012 following another flood event that occurred in March 2012.

The study used a mix of quantitative and qualitative methods. Anchoring on the quantitative sources is the household survey using a structured questionnaire providing the basis for a quantitative characterization of household socioeconomic background and recovery. A total of 97 households were purposively selected, calculated from the total number of households in the five selected villages. The sample size for the survey was determined using the sampling method devised by Arkin and Colton (1963) at 95% confidence level, with a precision level of ±4%. The sample size per village was proportionally determined based on the total number of households in the five villages (Table 1). Data was collected from the head of the households or any other elderly persons of the household who had detailed knowledge about the household experience in the two flood events and its recovery process.

For the qualitative part, semi-structured interviews were the appropriate choice. This exploratory study sought to catch "the complicated,

Table 1. Socioeconomic Background of the Surveyed Households.

Characteristics	Value	Characteristics	Value
Sample size HH surveyed in study villages		*Migration*	
Votua	33	Temporary (%)	35
Yalalevu	23	Permanent (%)	15
Matingara	18	*Education of household head*	
Singhawe	13	Illiterate (%)	14.4
Vatulaulau	10	Grade 1–8 (%)	48.5
Gender of HH surveyed		Grade 9–12 (%)	25.8
Male-headed household (%)	81.4	Above 9–12 (%)	11.3
Female-headed household (%)	18.6		
Ethnic composition of surveyed HH		Total agricultural land (mean in ha)	2.03
Indo-Fijians (%)	53.6	Use of fertilizer 2010 (mean kg)	5.2
Indigenous Fijians (%)	46.4	Use of fertilizer 2011 (mean kg)	3.4
Professions of HH		*Visits by extension agents*	
Farmers (%)	86.6	Yes (%)	45
Traders (%)	4.1	No (%)	55
Others (%)	9.3	*Access to credit*	
Family size (mean)	5.5	Yes (%)	37
Age of household head (mean)	53	No (%)	63
Dependent family members (mean)	2.4		
Economically active members (mean)	3.1		

HH, households.

ever-evolving and often conflicted feelings within individuals" and demanded a nuanced understanding of social capital access and use (Adler, 2009, p. 17). Personal interviews allowed us to go in-depth in context and look at the local dimension of social capital in a way that large-scale surveys neglect, an all-too-common oversight of social capital research today. Essays written by disaster-affected individuals were an additional tool that was used during the second fieldwork in Votua village. Notebooks with guiding questions were distributed to 23 female and 14 male villagers. They were asked to write how they were affected by and responded to the two consecutive floods in January and May 2012.

Participants in interviews were selected based on the recommendations of gatekeepers and key informants, as well as our own personal, often random social encounters. Only a few control variables were used in selecting respondents: age and gender, comparable damage sustained in the flood, and length of time in the area. Over the course of 14 days in Ba, we held 20 interviews of 45–70 minutes each, of which 11 were interviews with people affected, 7 were key informant and gatekeeper interviews, and 2 were pilot interviews. Ethics were appropriately considered in the design and implementation of this study. The chief ethical concern was the potential harm caused to respondents by discussing a sensitive, potentially traumatic subject (Adler, 2009). This concern proved nearly irrelevant, as residents were accustomed to discussing their flood experience repeatedly with outsiders.

SOCIOECONOMIC SITUATION OF SAMPLED HOUSEHOLDS

Households in the study area consist of both nuclear and extended families. Therefore, family members included parents, grandchildren, cousins, nephews and nieces. The majority of the households are, however, of the nuclear type.

Table 1 shows the major socioeconomic background of the surveyed households that are related to household decisions. The discussion in the remaining part of this section is partly based on this table. Note that percentages used in some parts of the discussion may not add to 100% when the interest is on the specific response rather than the distribution of the sample.

Economically dependent age groups (children 0–14 and elderly above 64) per household range between zero to three, with an average of 2.4

persons for the surveyed households. Most members of the dependent age group are children below 15 years of age. The number of economically active persons ranges from one in female-headed to eight in male-headed households. Of the total, 89% of the households have one to four economically active family members, of which one half are the parents. The average of the economically active persons per household is 3.1.

The livelihood of households in Ba District depends predominantly on farming, with the exception of four household heads who are traders and nine others involved in other professions. The majority of the farmers are involved in sugarcane cultivation, especially those with Indian ethnic background. Out-migration is 35% and 15% for temporary and permanent migrants respectively. Male temporary migrants total 75%, while 8.5% of the permanent migrants are female. Remittance income is obtained by 36% of the households.

CONTEXT OF THE STUDY: REPEATED FLOOD HAZARDS

The Ba District is located on the north-western side of Fiji's main island of Viti Levu (Fig. 1). The district is known for its severe socioeconomic problems related to continuing flooding. According to a provincial official, "floods have constituted a heavy drag on development of the district with a flood offsetting healthy economic growth for years." We found that floods had such a huge impact in Ba mainly because people were caught unaware and because of the rapid rate at which the water levels rise whenever there is a heavy downpour. Thus, many businesses suffered huge losses.

The sugar industry is the dominant activity in Ba District and the surrounding districts. Thousands of farmers in Ba and Tavua Districts supply cane to the Rarawai sugar mill of the Fiji Sugar Corporation located in Ba town. Testifying to the importance of the sugar sector as an employer in the area, the Ba Provincial report states "the importance of sugar industry in the life of Ba district may be gauged by the level of activity in Ba Town itself, which comes alive on the pay day of the cane harvesting season" (Ba Town Council, 2011). The district is known to be multicultural, mostly dominated by Indo-Fijians followed by Indigenous Fijians, with a few Chinese and European minorities.

In Ba floods are an annual occurrence like in many other places in the Western Division of Fiji. Often these floods are called "normal" by the

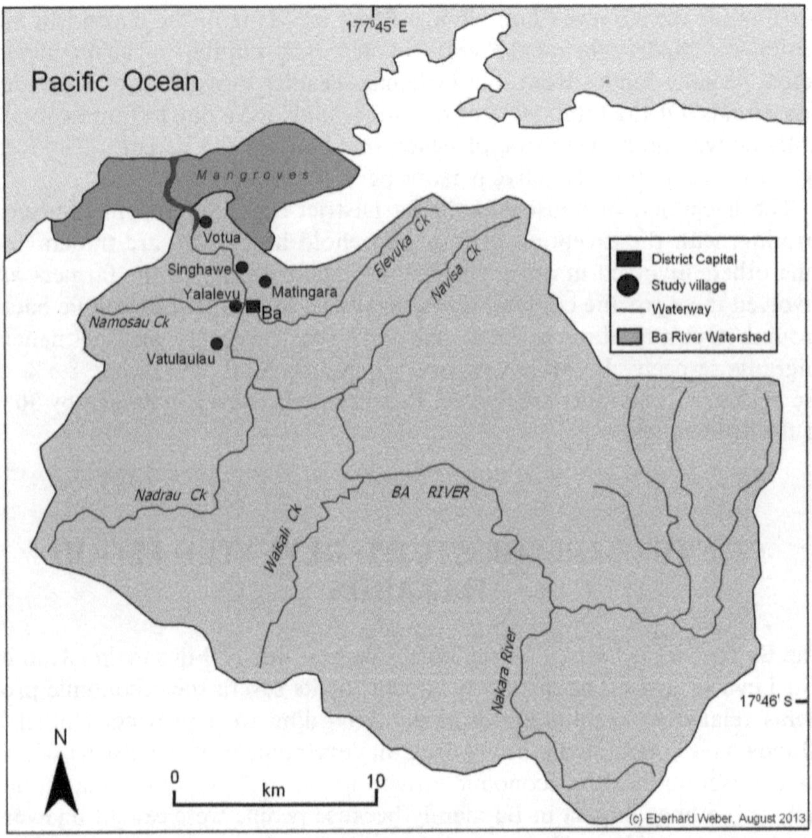

Fig. 1. Map of the Study Area.

people. They are looked upon by the residents as part of the natural landscape, and communities in the area have lived with the yearly challenge of flooding for generations. However, residents aptly noted that damaging floods, flood disasters, appear to have occurred with regular frequency in the recent years.

In the wake of the devastating floods of January 2009 and January 2012 which brought about much misery to many people of western Viti Levu, images of the tragedy and pleas for donations have captured world-wide attention. In 2012 the national government declared a "state of natural disaster emergency" in the flooded region. Flood disaster response was provided by governmental agencies, NGOs, FBOs and relief agencies, such as the Red Cross. In addition, active military personnel from all services were called to action at all levels and deployed through much of the affected areas for rescue, evacuation and recovery far exceeding those required for any previous disaster in the country. Private companies, such as Vodafone and Fiji Water, also provided flood relief.

Good cooperation was achieved between the military, multiple relief agencies, NGOs and FBOs which enabled resource sharing to victims of the flood disaster based on their capabilities. These efforts were well orchestrated through the regional government cluster meetings, which were held frequently to assess the status on meeting needs, such as food, shelter and sanitation.

As in past flood events, the January 2009 and January 2012 floods affected the most vulnerable groups. The majority of casualties were women and children. Farming communities sustained the largest economic losses due to the inundation of farm land and severe erosion. The survey showed that more than 75% of the households reported a substantial part of their livelihood lost to the two flood events. In comparing the devastation, a majority of people attributed that the 2009 was more devastating to their income, while the 2012 flood brought more silt resulting in more hours of gutting and cleaning their homes.

Although they were by no means the only floods in recent history, the short time lag between both events, combined with the magnitude of the events, resulted that the 2009 and 2012 floods were very much at the forefront of respondents' minds. Informal discussions and interviews with a wide cross-section of residents in the study areas highlighted comprehensive memories of the floods and their impacts on people, both directly and indirectly. The sudden occurrence of the 2009 and 2012 floods led to loss of life, erosion of farmland, disruption to transport and communications infrastructure, and also affected people directly in their homes (cf. Yeo, 2010). This immediate threat to people's personal survival and well-being perhaps explains why many household heads perceived floods to be a greater risk to their livelihoods and well-being than cyclones. One female head of household who lives close to the banks of the Ba River explained how they had to move out of their house well in advance before flood waters reached any threatening level during the two floods because they weren't sure how high the water levels would rise.

The households surveyed in the five villages had experienced and suffered severe floods in 2009 and 2012. About 36% and 24% of the households interviewed were forced to evacuate their homes at the peak of the floods in 2009 and 2012 respectively. More than 85% of these households reported to have taken shelter with extended families, neighbours and friends who provided a spare bedroom or cleaned out unoccupied space. They preferred such arrangements to seeking shelter in organized camps set by governments and relief agencies. Nearly 9% and 7% of the households had members that became ill, most commonly from intestinal complications and diarrhoea. The two successive flooding caused widespread losses to the agricultural economy and it was difficult for the villagers to recover swiftly from the damages. Although assistance from external agencies was helpful as reported by participants in the interviews, it was limited, compelling households to rely mainly on their own resources for rehabilitation. Reciprocal assistance from neighbours and friends was prominent at the onset of the floods and for a considerable time after the floods.

House damage was arguably the most fundamental and pressing impact of the floods. A much longer lasting consequence however has been the impact on agricultural land as it compromised livelihood generation for many farmers. In the Ba District farmland is situated along the river banks and within the Upper Ba Watershed. A large proportion of sugarcane production was lost during the floods. For one male-headed household the flood left a swamp in the middle of their plot, thus reducing the area available for planting. More usually, though, farmland was eroded. Of the households surveyed 25% had lost land in the past five years, each as a result of flooding. Some households lost the entire land they were cultivating. Others suffered substantial reductions: one of the few small-scale commercial farmers in Votua had his plot reduced almost by half, from five hectares to three. The loss of land also meant the loss of crops planted at the time of the flood reducing income from the respective sugarcane season.

Analysing the interviews it is surprising that few people talked about the impacts of the flood on agriculture. When asked they mainly referred to the physical impacts on their houses and land and psychological impacts. In fact, only 4% of the households surveyed mentioned losses of crops as a negative effect of the floods, although clearly a much higher percentage than that had damages to their standing crops. With regards to sugarcane production, the crop was still standing in the fields at the time of the floods. Harvesting usually starts early to middle of June. In 2012 the beginning of the crushing season of the Rarawai Sugar Mill in Ba was postponed to mid-July because of the flooding earlier in the year (The Fiji Times, 2012). One reason for not mentioning agricultural losses could be due to the perspective that losing crops is a minor negative impact as compared to the damage to houses and loss of land. Indigenous Fijians often have land in various places and crops destroyed near to the river banks usually is only part of their agricultural production (cf. Neef et al., Chapter 5, this volume). In the case of sugarcane production the sector has faced enormous difficulties for many years, and most farmers since long have not depended entirely on agricultural activities. They have other sources of livelihood and as such are in a better position to compensate damages to their crops standing in the fields. Indeed, there was a general expression that agricultural impacts were of lesser importance, partly as households have become accustomed to climatic variation and incidence of extreme events.

At the time we left the district, 30 days after the first 2012 flood (a second, even stronger one, followed at the end of March 2012), the rebuilding and recovery process from the January flood was still ongoing. Flood-affected residents continued to muck and gut their water-logged homes, unsure whether to stay or go. The local authority deliberated the merits of various flood-prevention proposals involving floodwalls, green space and river dredging. Meanwhile, residents waited nervously, wondering what the future would hold for the greater Ba region and its respective communities.

PATTERNS OF SOCIAL CAPITAL DEPLOYED TOWARDS RECOVERY

In the aftermath of the two devastating floods of 2009 and 2012 the stakes for deploying various elements of the social capital towards recovery could never have been higher. As in the case of many disasters, almost all the surveyed households decided to remain in the study district.

In this section, we describe the patterns households in Ba District adopted to make use of social capital within their communities. We identified four distinct strategies that were employed towards recovery. They are of particular relevance for how civil society might solve collective action challenges. These recovery patterns include practices of (1) search and rescue, (2) information dissemination, (3) mutual assistance and (4) socio-commercial cooperation.

Search and Rescue

At the onset of the floods expression of bonding and bridging social capital was readily evident. Respondents expressed how villagers helped themselves to evacuate people, organize patrol teams to guard and look after belongings from theft. Once the flood receded villagers and households worked together to repair community infrastructures such as canals, water gates and culverts. Most of this rehabilitation work was initially started by the villagers and then later aided by external agencies. This implies how people wanted to see their lives return back to normalcy as quickly as possible without having to wait for outside assistance, even at the expense of the deployment of individual and household resources.

The cumulative assessment of the impacts of the 2009 and 2012 floods showed that an average of 33% of the households reported to have taken refuge with friends and other families, 8% reported staying in temporary shelters, while only 5% of the households reported being rescued by volunteers. Although all households were affected as their homes were inundated during the two flood events, 51% stayed in their homes till the flood subsided. This implies the heavy reliance on social networks when it comes to rescue operations and living arrangements while flooding was ongoing. This calls for provision of rescue skills to residents in the flood-prone areas as component of preparation and recovery.

According to a study of the 2011 Thailand floods 76% of the victims were helped and checked upon by friends and neighbours (ADPC, 2012). The findings of our study reflect the same as much of the rescue operation during the two floods were coordinated by friends and neighbours. Only a few respondents stated that they were rescued by government and other agencies.

Participants in the interviews spoke highly of the efforts of the wider Ba District community in the aftermath of the flood, praising the helpfulness of villagers in the search and rescue efforts. Stirring examples of wider community involvement in the search and rescue were told to us frequently, usually with apparent gusto and pride. It seems that when floods strike the fault lines along the ethnic binary of Indo-Fijians and Indigenous Fijians in the context of Fiji dissolves, too. In spite of the heterogeneity of Yalalevu village, for instance, Rajendra – an Indo-Fijian – eludes that

> in the aftermath of a flood, the first help to arrive is often your neighbour – not a Red Cross agency, Sangam Fiji Foundation or donor country expert, but the helping hand of the person next door or a neighbourhood whether iTaukei (Indigenous Fijian) or Indo-Fijian. I think the search and rescue and general recovery has an opportunity perhaps to blur some of those boundaries.

Even when the Fiji military, Red Cross and experts arrived to fill gaps and meet needs, the local population still made up the large majority for search and rescue operations.

Information Dissemination

One of the strength of social networks is the facilitation of information as they serve as important channels for information. Our survey showed the indispensable role played by the social networks in facilitating flow of information in the flood affected areas. During the 2009 and 2012 floods, 57% and 61% respectively of the survey respondents reported that they learned about the areas set up for the victims to get relief assistance through the channels of friends, relatives, neighbours and their social ties (Fig. 2). Looking at the various information sources social networks were by far the most important source.

Another interesting aspect of the dissemination of information in the recent floods in Ba was the way some individuals were using social media to extend information, awareness and support. New applications of social and mobile media were also enabling new forms of response, creating new forms of social capital. In Fiji the use of social and mobile media picked up a lot in the two flooding events of 2009 and January 2012. It is apparent to note that rather than engage in reveries about the wonders of technology, there is a need to examine and focus on supporting the intersections and collaborations between these new forms of communication and long-standing ones such as radio and TV, which are still a valuable information source in flood crisis settings for many communities in Ba as shown in Fig. 2.

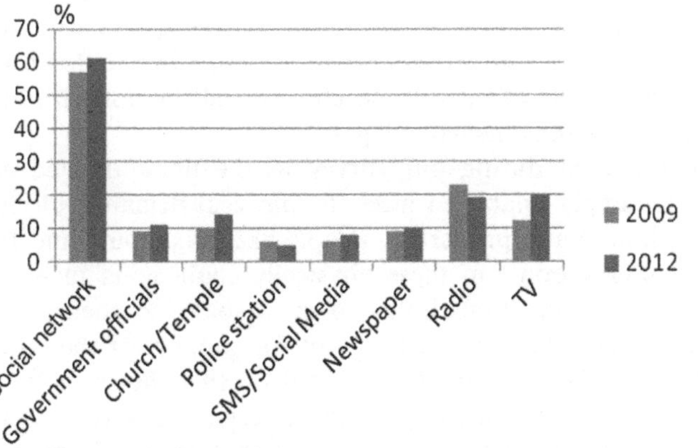

Fig. 2. Main Channels of Information Utilized by Households After the Floods.

We learned that those who took shelter with relatives, friends, or in the camps were more interested in hearing whether their neighbours had returned back to the village as many did not want to return to be the only ones in their village neighbourhood. The fear of becoming isolated and as a result exposed to a number of risks (including crime) was rather relevant. While the service providers were regularly providing updates on the status of electricity and water levels, such more individualized information was not (and probably could not be) reported.

Mutual Assistance

Mutual assistance emerges as one of the most prevalent aspect of social capital among respondents surveyed and interviewed. In contrast to charitable support which is a one-way form of help, mutual assistance is a strategy adopted by villagers to provide support to one another by short-term loans, free housing and shelter, tools and equipment, child care assistance and exchange of labour. Such support has reciprocal character and binds individuals and social groups' together, thus increasing cohesion of communities. As important as external aid and support was that came in towards recovery efforts, the small acts that people mutually offered at the time of the flood and its aftermath are what played distinct and vital roles in the recovery process – and beyond.

Mutual assistance was critical in initiating and guiding collective action in various ways. First, it served as source of material support among people. Second, it signalled that neighbours, friends and relatives are indisputably committed as partners in the recovery process. Third, it served in restoring the fabric of social networks torn apart by the flood which led to temporary relocation of people and constrained the normal functioning of community life in the affected villages.

As we observed during our survey in the flood-affected areas many people were simply unable to meet the physical demands of clearing debris in their homes and properties. People like neighbours and friends were willing to offer support in these physically challenging tasks. This direct support helped many households to move back to their houses sooner than if they would have needed to wait for support of the relief agencies. It also reduced the demands they placed on their hosts or for some who took temporary shelter in camps. Similar responses were observed when a social impact assessment was conducted in October 2009 in Samoa, a week after a devastating tsunami had struck the south-western coast of

Upolu (Gibson, 2010). Many relatives and friends of affected people came all the way from New Zealand after they learnt about this disaster to help in reconstruction and rehabilitation work. It is also interesting to note that in the last quarter of 2009 Samoa received the ever highest inflows of remittances from New Zealand, worth 47 million Samoan Tala (ST), compared to the earlier quarter when it was around 30 million ST. In 2009 it was around 9 million ST higher than the corresponding quarter of the previous year (Gibson, 2010).

Some of the respondents gave examples of bonding social capital during the floods. Mr Romolu, an Indigenous Fijian in his fifties of Vatulaulau village, described the benefits to his family from his social network during the January 2012 floods:

> We were overwhelmed with the support that came from near and far, friends, neighbours, family and acquaintances who were there for us the moment waters were rising, leading the evacuation of our household goods that night and following morning and their house a comforting refuge for us on those first shocking nights. What counted for me was that they turned up – and for those who physically couldn't come sent an SMS or phone messages to convey their thoughts.

However, the tasks of recovery are challenging when looked at in the light of rebuilding the complex fabric of human relationship that has been disrupted during the time of flooding. Considering the extent of damage wrought by the floods it can be difficult for households to recollect and start rebuilding except with clear assurance and signs that others are also willing and planning to do so. Mutual assistance therefore acts as indispensable signal and affirmation of others towards recovery and restoration of community.

Following the 2009 floods, Mr Luitimailagi, a Fijian resident of Votua village had his house inundated as well as the auto spare parts shop where he worked in Ba town. His Indo-Fijian employer provided a place for Luitimailagi and his family to stay in his compound for three months. This case illustrates that by giving and receiving mutual assistance, the two were affirming their willingness to share in the effort of recovery.

We noted that educational institutions played a key role in the recovery process and rebuilding of place-based social capital. Following the declaration of the state of natural disaster on 25 January 2012 by the national government for parts of the Western Division for a period of 15 days, schools in Ba District were forced to close for one week for safety reasons. Several school buildings in the affected communities were used as evacuation centres for flood victims (Figs. 3(a) and (b)).

Fig. 3. (a, b) School in Votua Serving as Evacuation Centre During the January 2012 Flood. *Source:* Courtesy of Votua Community.

Parents had a strong desire to bring a sense of normal life back to their children by opening flood-affected schools as early as possible. Many parents and school officials took time alongside teachers and students to clean up the schools and salvage what was left. This mutual assistance offered a swift and efficient way of bringing back on track the educational life as well as created opportunity for shared memories and points of contact between young people and adults. All these experiences will help to form a greater sense of connection for the community in the long run.

Socio-Commercial Cooperation

Discussions on social capital often miss out on market activity. Whether this is by omission or commission is out of the scope of this chapter. In our study we considered the role that commercial activity played in the post-flood recovery in the study area. At an individual level, commercial activities were important to sustain families' livelihood while waiting for outside assistance as reflected in this quote from one of the written essays:

> After the flood, I looked for some sources of income available such as selling crabs and prawns and fish, while waiting for the government support and other donation from private, religious and individuals. (Vasiti, young woman from Votua)

Beyond the individual level, socio-commercial cooperation was crucial in the recovery process. As pointed out by Chamlee-Wright (2010, p. 22), "commercial cooperation puts business activity at the centre of the disaster recovery." For profit making, commercial cooperation in many ways serves a similar function as mutual assistance through the provision of essential material support, builds confidence about rebuilding of community, and impetus for redeployment of place-based social capital.

It is worth noting that material support offered by neighbours, relatives and friends finds more meaning if complemented with necessary equipment, tools and building materials which becomes available through business activity. The early reopening of hardware stores in the flood areas enabled many households to undertake the rehabilitation of damaged houses as quickly as possible. Given that the majority of the homes in the area were not insured in the event of floods or storms most households rehabilitated the destruction on their houses by themselves. Without the reopening of these hardware stores providing low-cost, easily accessible materials, it would have been difficult for victims of the flood to rebuild and even gain collective action from neighbours and friends to assist.

Appreciating that their fate were closely linked, some businesses were willing to make advance payment to their workers and extend credits to their customers even with the knowledge that payment may be significantly delayed. The immediate reopening of commercial activities served as effective signal that the Ba community was rebounding towards recovery. In order to recover from the floods, people needed to return back to their homes. In the study, among the most frequently cited frustration was the delay in the restarting of grocery stores. The reasons are rather obvious as many of the farms were damaged by flooding and shops' inventories were destroyed. It took a while before shops had been restocked through goods brought in from other parts of Fiji and overseas. The re-establishing of commercial life is one of the most important aspects of recovery, which is also considered by many emergency plans. Our findings suggest that it also renders social capital more effective.

It is important to point out that commercial activities do not serve merely as a cheerleader for social capital, but also as principal provider of social capital. For instance, *Yagona* cafés, tea shops, restaurants and bars served as joints for reconnection after the floods and provide the social spaces where residents reminisce about the flood and exchange practical advices. The conversation that goes over the *tanoa* and cups of tea offer people a reprieve from the cleaning, gutting and rebuilding as well as cushion the effect of the flood tragedy.

Many authors acknowledge the vital role of social capital in the smooth functioning of markets. The norms of generalized trust and reciprocity explain how market exchange lubricate and reduce transactions costs and associated uncertainty (Chamlee-Wright, 2010; Fukuyama, 1995, 1999; Putnam, 1993). However, what is often given less prominence and attention in literature is the importance of commercial activity with regards to the development of social capital. The accounts of the flood victims in Ba District – be they village leaders (*turaga ni koro),* teachers, business owners, religious leaders or farmers – suggest that the social capital redevelopment is directly and intimately tied upon the market redevelopment. Under the condition of post-flood recovery as seen in Ba, commercial cooperation played a significant role in the recovery process by offering the residents the opportunity to orchestrate successful recovery and perspectives on how communities may thrive once again. Hence, we find that there is a dialectic relationship between social capital and disaster recovery: disaster recovery does not just depend on the existing social capital in a community, but the recovery process itself plays a major role in reinforcing bonding social capital. This is further discussed in the following section.

DISASTER RECOVERY AS FACTOR OF REINFORCING BONDING SOCIAL CAPITAL AND BUILDING BRIDGING SOCIAL CAPITAL

Many of the residents believed that the flood had brought about the feeling of togetherness and norms of helpfulness and trust to Ba. They overwhelmingly focused on the fact that the flood affected everyone in some way, directly or indirectly, and consequently had a unifying effect. The waters, they point out, did not discern between victims and businesses. All people suffered together "in the field" rather than "in the palace," analogized Pastor Tukana. Differences of colour and creed blurred, as it became "vividly clear that we all are one when it comes to human needs of shelter, food, hopes and the dreams." Alanieta agreed, noticing a change in the district after the flood: "When you go to the department stores or out in the floodway, people were affected, whether they were directly affected or indirectly." And our community was like, "wow. And so there were a lot of people – strangers – hugging you." Alipere saw the same phenomenon happening while waiting in long lines, trying to get some relief assistance: "As much as you hated standing in lines, you had to break the boredom by talking." From these exchanges, "you'd almost feel like you knew them because they didn't live that far away and they had their own story, and it made you feel a little bit better because you knew you weren't alone."

Norms regulating interaction eased – it became acceptable to just strike up a conversation with anyone if it began by concentrating on the flood.

Adler (2009, p. 23) writes about the significance of these exchanges: "Because sharing stories necessarily involves others to hear them, the process helps to reconnect victims, who may feel alone and alienated, to one another as they form a collective identity. The sharing of narratives also helps a community to gather individual experiences together to construct a mosaic of shared meanings of the flood disaster." Through these casual but purposeful conversations, social networks grew and the sense of togetherness in Ba increased.

For others, flood-centred conversations yielded valuable information that led to additional financial support, help from volunteers, and even future work contacts. "The relationships I've built working with people to find solutions, whether it be in my job or neighbourhood, it would have taken ten to fifteen years or longer to establish those relationships. So when I look down the line in the future...you know: are you really better and stronger from having this happening? Yes," beamed Pillay.

As individual and societal needs merged, an ethos of helpfulness took centre stage and the community united. "People just kind of joined together," recalled Rusila. "In Ba district, because we have lost so much, people have really worked together," said Chand. Pamela also observed the cooperative spirit of residents: "everybody seemed to pitch in and offer whatever was necessary [...] that's just what you do." "When things were at their worst [...] everyone was pulling in the same direction," remarked Shameet.

What is the effect of this coming together? "There are blessings to be had," said Alanieta. "I think what it does is: people step back and they go 'well I have a lot; now I need to share.' And I think the community of Ba has really done well with that. I really do." Kirti is also optimistic. She predicts community life will be even better than before in Ba, more robust and with a more acute sense of civic pride after these two subsequent floods. When asked about the reason, she said, "we overcame this together, and we can do it." Because of the far-reaching devastation of the flood and the initial coming together of residents, the flood is a positive integration landmark that has "affected the hearts and minds" of all residents, concluded Pastor Tukana.

Aside from reinforcing bonding capital, the two consecutive flood disasters in January and March 2012 also contributed to building bridging social capital, i.e. bonds between local communities and outsiders, such as government agencies, NGOs and private companies. This was evident from the follow-up inquiry in September 2012 that was conducted in Votua community after the second flood of 2012. Many respondents mentioned in their written essays that they were grateful for the outpour of support from external organizations, such as the Red Cross Fiji, the military, government agencies, religious organizations, the international donor community, and even private companies as part of their corporate social responsibility (CSR) activities. Government assistance was most often mentioned in the 37 essays (21 occurrences), followed by support from NGOs (17), assistance from faith-based organizations (10) and help from domestic private companies (7). The following three quotes exemplify typical responses:

> I received aid and financial assistance donated by religious bodies which were very helpful in addressing our needs. Fresh water was donated by the Fiji Mineral Water company. The Red Cross donated clothes, food parcel and bedding materials. (Maikeli, farmer from Votua)

> A rehabilitation program was conducted by the government in rebuilding infrastructure, such as roads and bridges. At the same time, they provided food rations for every household to help them slowly restore their normal livelihood. (Makitalena, housewife from Votua)

> Fortunately, government and private firms came to our assistance, such as Fiji Water Ltd. supplied us with drinking water, Hare Krishna with food parcels, Red Cross with clothes and food packages and some churches supplied other important things to us in order to rehabilitate ourselves and the whole village. The fire brigade came to help clean up the school and its premises with some military personnel who helped us in cleaning the debris and the village as a whole. (Mareta, female youth from Votua)

As the last quote suggests, these external organizations oftentimes joined hands with local people in well-coordinated rescue and recovery efforts (cf. Figs. 4(a) and (b)). Such joint activities established new forms of bridging social capital. Yet not all Votua communities shared this positive

(a) (b)

Fig. 4. (a, b) Joint Disaster Recovery Efforts by Red Cross Staff and Votua Villagers. *Source:* Courtesy of Votua Community.

perspective; while one Indo-Fijian family expressed their gratitude for having received donations from international aid agencies, in another Indo-Fijian community people told us in an interview that they were bypassed by external assistance, because they were deemed less affected than their indigenous Fijian counterparts and therefore were supposed to be able to cope with the flood without help from outside.

DISCUSSION, LIMITATIONS AND SUGGESTIONS FOR FURTHER RESEARCH

This chapter explored a largely unexplored link between two well-developed fields in the context of very recent flood events. Our research is not necessarily representative of other circumstances. Ba District is a relatively heterogeneous mid-sized district with high levels of poverty and without heavy social conflict. The district is especially typical for the Western division of Fiji, though no district can be taken as a microcosm for wider society. The goal of this study was to understand, in nuanced terms, how two particular flood events impacted an outlined version of social capital in a district through the perceptions of its residents.

Yet, the recovery is far from over, and many residents agreed that the post-traumatic stress and despair are just setting in: "we are *only* hitting the tip of the iceberg in acknowledging our own pain," said the Singahawe village leader. So, while the research respondents did share their experiences before and following the floods, they may have overlooked or omitted some important details. Furthermore, under these conditions research respondents may have been under heavy stress, which makes their perceptions likely to be skewed or memories becoming fallible. In short, much more contextual research is needed to obtain a thoroughly detailed understanding of the flood and its effect on social capital. However, the findings are important in the sense that respondents expressed their personal experiences rather than factual representation of the flood events.

As our case study indicates, there is a distinct need for a research approach mindful of extensive nuance when studying social capital and disasters. As a field of inquiry, social capital would be well-served to reduce its reliance on the top-down approach of survey measurement and opt instead for more bottom-up, contextual understandings rooted in qualitative community case studies (Adler, 2009). In addition, the metrics in gauging recovery are speed and quality. However, this study only looked at the speed of recovery and did not address the quality of recovery per se. Hence, future studies should focus on this aspect of recovery.

Flood disaster studies should be incorporated into future social capital research, and vice versa, narrowing the curious gap between canons. The overlap is apparent: both fields seek to understand community functionality and identify inequalities within complex social systems. The study of social capital is a study of structural and cultural resource advantages between individuals and within societies. The study of disasters is a study of human and societal vulnerabilities as exposed by natural hazards. A more interdisciplinary future between both fields stands to offer each key insight on its own fundamental concerns. As this chapter illustrates, the need to synthesize disaster studies and social capital is conspicuous: together, they provide an opportunity to study normally latent social processes that help us understand the structural and cultural order of a community.

CONCLUSION

It is evident from the findings that social capital is an important resource with the capacity to complement government flood recovery efforts. Communities in flood-prone Ba District relied much upon existing complex and decentralized social structures and processes to leverage upon towards their recovery.

It was clear that in the early phase of flooding much of the search and rescue endeavours came from within the social networks, likewise in the aftermath social networks facilitated significant levels of flow of information, provision of different types of support both in kind, financial, material and emotional as well as enhancing commercial cooperation.

Given the important role of social resources, it is pertinent to ask what should be done in terms of future flood policies in Fiji. First, there is need for proper acknowledgement of the critical role of social capital and social resources by policy-makers and the non-profit sectors as often social capital and existing local networking are just mentioned peripherally in disaster planning, if they are mentioned at all. There should be more active deliberations of how to strengthen such networks. In spite of the robust contribution of social networks, what is often lacking – and what government, NGO and international actors can and should support – is capacity. It is indisputable that society's flood disaster recovery depends heavily on the immediate capacity of the local community to respond with the appropriate skills, accountability, and technical expertise. The effectiveness of responses to future disasters will depend on local capacity and resilience.

As observed, most money flowing into flood disaster response in Western Viti Levu is restricted to providing food, shelter, health care and other immediate services, not building the capacity of local groups. This is in large part due to government's and many donors' emphasis on the delivery of services and the media's attention on immediate results. When solicited, feedback from the local community often highlights the lack of capacity-building and disaster risk reduction efforts. In this regard, government agencies and NGOs should try to make full use of existing social capital in recovery by maintaining social networks in disaster-affected areas and utilizing the existing social networks in recovery processes. Government bodies and NGOs can consider the provision of communications devices such as cell phones and e-mail connections to community groups which can facilitate them to stay in contact during the flood situation, as well as sponsor information sessions targeting the established neighbourhood-based or faith-based groups. They should also consider putting displaced people from the same community together, as opposed to evacuating them to different locations, as this has far-reaching positive consequences for communities struck by flood hazards as it helps to maintain and sustain the sense of community. However, it is important to realize that investment in social networks/capital is a long-term one, as opposed to emergency policy response.

Lastly, various public policy programmes have the capacity to build up stores of interaction and trust among members of a neighbourhood. In the aftermath of the Kobe earthquake in Japan, the city worked on creating a strong solidarity among the survivors through enabling programmes that built increased trust and participation of residents. Similar programmes can be replicated for the flood-prone Western Viti Levu. Thus, social capital could be generated through implementation of proactive policies.

ACKNOWLEDGEMENTS

Special thanks to the residents in Ba District who spared their time to participate in this study amidst their trauma of recovering from the January 2012 flood. Villagers in Votua community kindly shared their written essays and their photos during the follow-up fieldwork in September 2012.

Funding for the study was provided through a grant from the Faculty of Science, Technology and Environment of the University of the South Pacific (USP), Fiji, and through a grant-in-aid by the Japan Society for Promotion of Science (JSPS) for the research project "Integrated Assessment of Post-Disaster Recovery in Coastal Communities in Asia-Pacific" (JSPS KAKENHI Grant Number 24401007).

REFERENCES

Adler, K. F. (2009). The structural and cultural impact of natural disasters on social capital: exploring community life after the 2008 flood in Cedar Rapids, Iowa. Conference Paper presented at the Midwest Political Science Association 67th Annual National Conference, Chicago, April 2–5, 2009.

ADPC (Asian Disaster Preparedness Center). (2012). *Lessons on resilience from the 2011 Central Thailand floods*. Bangkok, Thailand: ADPC.

Aldrich, D. P. (2010). Fixing recovery: Social capital in post-crisis resilience. *Journal of Homeland Security, 6*, 1–10.

Aldrich, D. P. (2011). The externalities of strong social capital: Post-tsunami recovery in southeast India. *Journal of Civil Society, 7*(1), 81–99.

Appadurai, A. (1984). How moral is the South Asian economy? A review article. *Journal of Asian Studies, 43*(3), 481–497.

Arkin, H., & Colton, R. (1963). *Table for statistics* (2nd ed.), New York, NY: Barnes and Noble.

Boettke, P. J., Chamlee-Wright, E., Gordon, P., Ikeda, S., Leeson, P. T., & Sobel, R. (2007). The political, economic, and social aspects of Katrina. *Southern Economics Journal, 74*(2), 363–376.

Bourdieu, P. (1986). The forms of capital. In J. Richardson (Ed.), *Handbook of theory and research for the sociology of education* (pp. 241–258). New York, NY: Greenwood.

Chamlee-Wright, E. (2010). *After the storm: Social capital regrouping in the wake of Hurricane Katrina*. Global Prosperity Working Paper No. 70. George Mason University.

Coleman, J. S. (1988). Social capital in the creation of human capital. The *American Journal of Sociology, 94*, 95–120.

Coleman, J. (1990). *Foundations of social theory*. Cambridge, MA: Harvard University Press.

Duaibe, K. (2008). *Human activities and flood hazards and risks in the South West Pacific: A case study of the Navua Catchment area, Fiji Islands*. Masters thesis, Victoria University of Wellington.

Fukuyama, F. (1995). Social capital and the global economy. *Foreign Affairs, 74*, 89.

Fukuyama, F. (1999). *The great disruption. Human nature and the reconstitution of social order*. London: Profile Books.

Gibson, J. (2010). Recent shocks and long-term change in the Samoan economy. *Pacific Economic Bulletin, 25*(3), 7–23.

Hawkins, R. L., & Maurer, K. (2010). Bonding, bridging and linking: How social capital operated in New Orleans following Hurricane Katrina. *British Journal of Social Work, 40*, 1777–1793.

Mudgway, L. (1997). Watershed management and flood control on Viti Levu, Fiji. In A. Planitz & J. Chung (Eds.), Proceedings of the disasters and development session, VIII Pacific science inter-congress, 13–19 July 1997. The University of the South Pacific, Suva, UNDHA – SPPO, Suva (pp. 104–113).

Putnam, R. (1993). *Making democracy work: Civic traditions in modern Italy*. Princeton, NJ: Princeton University Press.

Putnam, R. D. (2000). *Bowling alone: The collapse and revival of American community*. New York, NY: Simon and Schuster.

Sami, Y., & Reddy, M. (2011). *Beyond neo-classical economics: Disintegration of social capital and trust in an agrarian economy*. Suva, Fiji Islands. Oceania Development Network.

Sano, Y. (2008). The role of social capital in a common property resource system in coastal areas: A case study of community-based coastal resource management in Fiji. *SPC Traditional Marine Resource Management and Knowledge Information Bulletin*, 24, December.

Schuller, T., Baron, S., & Field, J. (2000). Social capital: A review and critic. In S. Baron, J. Field, & T. Schuller (Eds.), *Social capital* (pp. 1–38). Oxford: Oxford University Press.

Serra, R. (2001). Social capital: Meaningful and measurable at the state level? *Economic and Political Weekly*, *36*(8), 693–704.

SOPAC. (2002). *Regional comprehensive hazard and risk management: Guidelines for Pacific Island countries*. Suva: Pacific Applied Geosciences' Commission.

SOPAC. (2009). *River floods. Natural hazards in the Pacific fact sheet*. Suva: Pacific Applied Geosciences Commission.

Sopha, A., Oeur, I., & McAndrew, J. (2007). Understanding social capital in response to floods and droughts. *Cambodia Development Review*, *11*(4), 9–12.

Szreter, S., & Woolcock, M. (2004). Health by association? Social capital, social theory, and the political economy of public health. *International Journal of Epidemiology*, *33*(4), 650–657.

Terry, J. P., McGree, S., & Raj, R. (2004). The exceptional flooding on Vanua Levu Island, Fiji, during tropical cyclone Ami in January 2003. *Journal of Natural Disaster Science*, *26*, 27–36.

The Fiji Times (Canada). (2012). Sugar crushing to begin this month. *The Fiji Times*, p. 17. Retrieved from http://de.scribd.com/doc/97844433/FijiTimes-June-22-2012-PDF

Veitayaki, J., Aalbersberg B., Tawake, A. Rupeni E., & Tabunakawai K. (2003). *Mainstreaming resource conservation: The Fiji Locally Managed Marine Area Network and its influence on national policy development*. Technical Report Working Paper No. 42, Resource Management in Asia-Pacific (RMAP) Program. Research School of Pacific and Asian Studies, Australian National University, Canberra.

Yeo, S., Blong, R. J., & McAneney, K. J. (2007). Flooding in Fiji: Findings from a 100-year historical series. *Hydrological Sciences*, *52*, 12.

Yeo, S. W. (2000). *Ba community flood preparedness project: Final report*. SOPAC Technical Report 309.

Yeo, S. W. (2010). Fiji's worst natural disaster: The 1931 hurricane and flood. *Disasters*, *34*(3), 657–683.

Zhao, Y. (2010). *Social networks and reduction of risk in natural disaster: An example of Wenchuan earthquake*. Switzerland: Global Risk Forum GRF Davos.

CHAPTER 5

THE INTERPLAY BETWEEN COLLECTIVE ACTION, INDIVIDUAL STRATEGIES AND STATE INTERVENTION IN MITIGATING FLOOD DISASTERS IN THE UPLANDS OF NORTH THAILAND AND NORTHWEST VIETNAM

Andreas Neef, Peter Elstner and Iven Schad

ABSTRACT

Drawing on studies in flood-affected upland areas of Thailand and Vietnam, this chapter explores the complex interplay between collective, state and individual responses to disastrous flood events and subsequent mitigation strategies. Fieldwork was conducted between 2007 and 2009, employing a variety of qualitative methods, such as semi-structured interviews in flood-affected households, focus group discussions and narrative

Risks and Conflicts: Local Responses to Natural Disasters
Community, Environment and Disaster Risk Management, Volume 14, 109–130

ISSN: 2040-7262/doi:10.1108/S2040-7262(2013)0000014011

essays written by local people. Evidence suggests that farmers' willingness to engage in flood mitigation is curbed by the common perception that flooding is caused by a bundle of exogenous factors. In the case study from Vietnam, state intervention in formerly community-based water management has alienated farmers from water governance and reduced their sense of personal and collective responsibility. Their lack of engagement in flood-prevention strategies could also be explained by the fact that their major cash crop was not affected by the flood event. In the Thai case study, where community-based water management remained largely unaffected by state influence, villagers agreed in a collective decision-making process to widening the riverbed after a severe flood, although this meant that some farmers had to give up parts of their paddy fields. Yet, following a second flood, these farmers opened up new upland rice fields in the forested upper watershed areas to ensure their food security, thus increasing the likelihood of future flood disasters downstream. We conclude that flood mitigation and adaptation policies need to consider (1) local people's own causal explanations of flood events and (2) the potential trade-offs between collective action, state intervention and individual livelihood strategies.

Keywords: Mitigation; adaptation; livelihood strategies; flood disaster; Southeast Asia

INTRODUCTION

Lowland areas of Thailand and Vietnam have experienced an ever higher frequency of floods in recent years, mostly ascribed to land use changes in upstream watersheds and to rapid urbanisation (e.g. Lebel, Manuta, & Garden, 2011; Tran, Marincioni, & Shaw, 2008a). Floods in mountain watersheds have received much less attention both in the media and among the scientific community, as their impact tends to be more localised (Schad et al., 2012). Hence, mitigation of and adaptation to floods in mountainous environments of mainland Southeast Asia are still poorly understood.

Among the Thai ethnic groups in the uplands of North Thailand and in Northwest Vietnam, communal water resource management has a long tradition and has been described as an efficient, sustainable and participatory system of water allocation. These Thai groups, such as the Shan and northern Thai (*khon muang*) in the Thai highlands and the Black Thai in the northwestern mountains of Vietnam, have common ancestors in Southwest China who migrated southwards several centuries ago to establish powerful states and principalities in various parts of upland Southeast Asia (cf. Wyatt, 1984). Wherever they found areas suitable for wet-rice cultivation, they began to develop highly sophisticated water management schemes, many of which have survived until today (Neef, Chamsai, & Sangkapitux, 2006). These systems, locally known as *muang-fai*, have been the focus of several studies (e.g. Cohen & Pearson, 1998; Ounvichit, Wattayu, & Satoh, 2008; Sirivongs Na Ayudhaya, 1979; Surarerks, 1986, 1998; Tan-Kim-Yong, 1995; Tang, 1992). However, most studies focused on a rather static description and tended to neglect the dynamics of these systems in different socio-political contexts and under the perils of climate change.

Scholars working on the social dimensions of flood events, public perceptions of their causes and impacts and the resulting mitigation and adaptation strategies have focused either on perceptions, impact and responses at the community level or at the individual household level or on the success or failure of state intervention (e.g. Brilly & Polic, 2006; Heitz, Spaeter, Auzet, & Glatron, 2009; Lebel et al., 2011; Lin, Shaw, & Ho, 2008; Paripurno, 2006; Tran, Marincioni, Shaw, Sarti, & An, 2008b). In this chapter, we argue that significant trade-offs may emerge from the interplay between collective action, individual strategies and government intervention towards mitigation of flood events and adaptation to flood disasters. This may have a strong bearing on the potential of implementing integrated flood prevention and adaptation programs.

The remainder of this chapter is organised as follows: in the next section, we briefly describe the study areas and research methodology. We then present the research findings from the two case studies in North Thailand and Northwest Vietnam, looking at the impact of the floods on communities and households, the immediate responses by villagers, communities and external relief agencies, the causal explanations of the flood events, and long-term adaptation and mitigation strategies. We then synthesise and discuss the findings and draw conclusions.

STUDY AREAS AND METHODOLOGY

The location of the two study areas, the Mae Lana Watershed in North Thailand and the Chieng Khoi Watershed in Northwest Vietnam is depicted in Fig. 1.

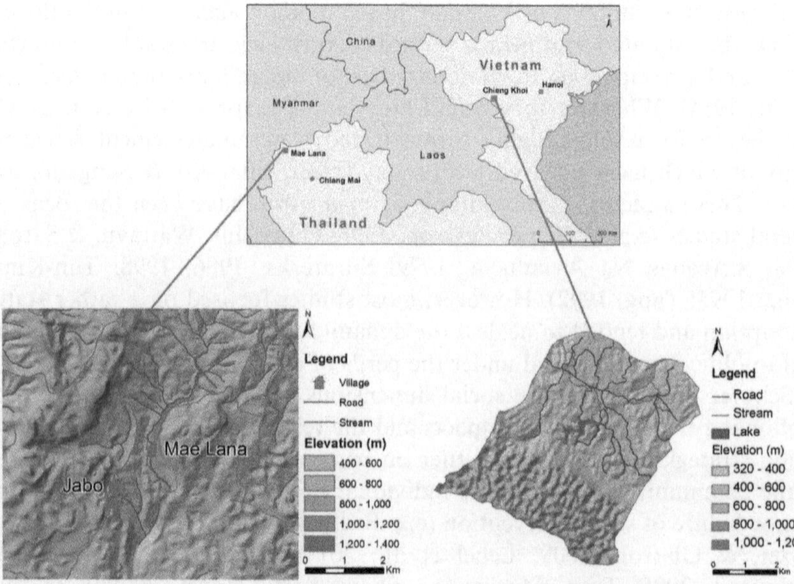

Fig. 1. Location of the Mae Lana Watershed, Pang Ma Pha District, Thailand (Left) and ChiengKhoi Watershed, Yen Chau District, Vietnam (Lower Right).

Study Area in North Thailand: The Mae Lana Watershed

The Mae Lana Watershed is located in Pang Ma Pha district, Mae Hong Son province. The mountainous landscape is characterised by limestone, sandstone and volcanic rocks. The vegetation is dominated by mixed deciduous forest and some patches of dry evergreen forest. The average annual rainfall is 1,300 mm. The Shan village Mae Lana is located in the valley of the Mae Lana River and is the oldest settlement in the sub-district with a history of more than 200 years. The plain valley area is classified as suitable for permanent agriculture and thus the farmers have official land use rights. The surrounding mountains are classified as watershed class 1, which implies that this area should be maintained as protection forest with no agricultural cultivation being allowed. Mae Lana has limited access to the market with partly earthen roads and a long distance to the provincial capital. As a consequence, paddy rice production for subsistence is the main agricultural activity. Since the irrigated land is not sufficient, upland fields are cultivated with upland rice, maize, sesame and fruit trees.

Study Area in Northwest Vietnam: The Chieng Khoi Watershed

The Chieng Khoi Watershed is part of the Chieng Khoi commune, Yen Chau district, Son La province in the northwestern uplands of Vietnam. It comprises an area of about 800 ha and is located on a mid-level plateau between a steep mountain range and the valley of Yen Chau. The average annual precipitation is 1,200 mm, with the rainy season generally lasting from April to September. The flat plateau is used for paddy rice cultivation – divided into a spring and summer rice crop – and aquaculture, while the surrounding upland fields which account for more than 85% of the total land area are planted with maize and cassava as main crops. Black Thai settlements in this area have a history of several hundred years. The land allocation process was completed in 1999, when all households had received a share of agricultural land according to land category (i.e. upland fields, paddy fields, forestland, fishponds), social status and number of family members. Nearly all households own at least one fishpond, providing an important source of protein for human consumption.

Data Collection and Analysis

Data for the two case studies were collected through in-depth qualitative fieldwork conducted between 2007 and 2009 and built on earlier research on community-based water management in the two study areas (cf. Hager, 2005; Neef, Elstner, & Hager, 2006).

In the study in North Thailand, open and semi-structured interviews with key informants (e.g. village headmen, government officials, heads of water management committees and women groups) provided insights into customary irrigation management and various stakeholders' responses to the flood events in 2005 and 2006. Field notes were immediately discussed and cross-checked between the German researcher and the Thai research assistant after the interviews. The location of the major components of the irrigation infrastructure system, such as canals and weirs, was determined with a mobile Global Position System (GPS) and integrated into a Geographic Information System (GIS).

Data collection for the study in Northwest Vietnam involved semi-structured interviews, focus group discussions, ranking exercises and risk mapping. In the second step, selected members of flood-affected households and key representatives of the village management board were asked to write narrative essays on various aspects of the flood, such as their own causal explanations for the flood event, the impacts of the flood on their family's economic and social situation they regarded as important. This information was triangulated with data obtained from the individual interviews and focus groups discussions. All events were recorded, translated into English and transcribed, followed by computer-assisted qualitative data analysis. The social science component was complemented by data from two natural science projects (on local hydrology and nutrient flows) and a socio-economic project, but these are not in the focus of this chapter.[1]

RESPONSES TO REPEATED FLOOD EVENTS IN THE MAE LANA CATCHMENT, NORTH THAILAND

Management of Agricultural Land and Irrigation Water Prior to the Floods

In total, around 25% of the households in Mae Lana own land suitable for paddy rice cultivation and possess registered land use rights. These villagers are mainly descendants of the first settlers. Selling of this land is not common as it is the main asset for these families and the basis of their high social and economic status. The land is controlled by the head of the family clan and cultivated by different families of the clan according to their needs and labour capacities. Households that do not belong to the long-established families cultivate upland rainfed fields without any legal land rights or depend on leasing contracts of paddy fields. In this sharecropping arrangement – under which they have to give 50% of the harvest to the landowner – the tenant receives not only the right to use the land but also the right to share the water from the communal irrigation system described above. Thus, land and water rights are closely intertwined as only farmers – whether owners or sharecroppers – with fields in the plain area along the canals can get access to irrigation water from this system. Access to water also depends on the location of the fields. Paddy fields located at a higher altitude than the canal do not receive water as farmers cannot afford to purchase water-lifting devices.

The only water source for irrigation in the Shan village Mae Lana is the river of the same name. It has its headwater near the border to Myanmar and flows along the valley passing several villages before disappearing in a cave at the end of the valley in Mae Lana. Water management is based on the traditional communal *muang-fai* system[2] which has existed for centuries. Paddy rice as the main crop is cultivated only in the rainy season. In the dry season some farmers in the upper part of the village territory cultivate soybean, onion and garlic.

Mae Lana has three weirs built from concrete which replaced the old constructions made of bamboo and stones (see Fig. 2). For each weir, a different government agency provided the funds. The canals leading from the weirs to the fields are also made from concrete and were constructed with support from the Royal Irrigation Department and the Local Administration Office. Despite the intervention of numerous external organisations in providing the irrigation infrastructure, the management of water is entirely under the control of the community. Weir 1 and the respective canal constitute the only system in the village managed by the *gae muang*, an elected irrigation manager. The responsibilities of the *gae muang* are to (1) ensure an equitable water allocation during the rainy season, (2) organize and supervise the communal cleaning of the weir and

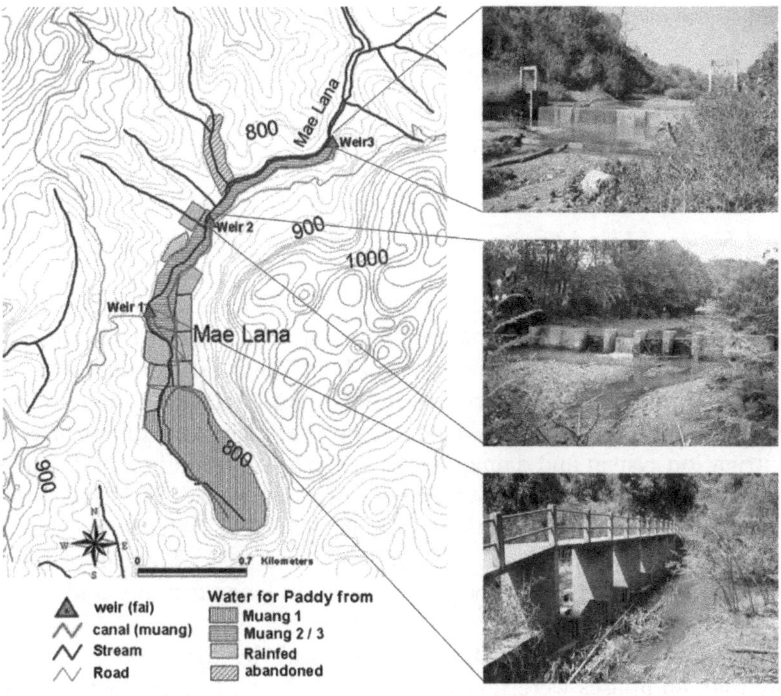

Fig. 2. Irrigation System of Mae Lana. (Map and Photos by P. Elstner).

canal, which is done twice a year and (3) to perform the ceremony for the water spirit prior to the irrigation season at the headwater of the Mae Lana river to ask for abundance of water and protection of the crops.

At the time of the flood event, the *gae muang* had been in charge for 10 years. He received a specified share of the rice harvest as compensation for his work. He had been an assistant of the former *gae muang* and was elected when his predecessor retired which points to the continuity of the management of the system.

Weir 2 was built by the Local Administration office in the mid-1990s and is of a simple structure. Only the sides of the weir are constructed from concrete and the villagers use bamboo stakes in the centre to divert water to the canal. Weir 3 was constructed by the Accelerated Rural Development Unit at the same time as Weir 1. The canal leads from Weir 3 partly underground to Weir 2 where both canals join and then further along the street to the village. Weirs 2 and 3 provide water for the narrow valley north of the village for the fields of 15 farmers (see Fig. 2 above). These farmers manage the system themselves by working in groups. They organise the maintenance and cleaning of the weir and canals as communal work, and each farmer allocates water to the fields as required. The farmers in this area would start the cultivation before the users of Weir 1. As they are the first users in Mae Lana, water is also sufficient to allow some farmers to grow crops during the dry season.

A Series of Flood Events and their Impacts on Villagers' Livelihoods

In the year 2005, the village Mae Lana was hit by a series of floods, two of them being severe. The first flood occurred on the morning of 13 August. A weather station installed in the nearby village Bor Krai recorded a rainfall of 161 mm on that day. This event caused minor damages in the residential area and flooded some paddy fields along the river, yet without affecting the irrigation system and damaging the rice crop. A second more disastrous flash-flood happened on 28 September, after seven days of heavy rainfall and an amount of 70 mm on that particular day. The water level in the village began to rise slowly in the early morning and reached a peak of 150 cm at around 10 am, making it the biggest flood in living memory and causing havoc in the community (Figs. 3(a) and (b)).

Four houses were completely washed away by the flood waters and more than 40 houses were severely damaged. Many farmers lost their pigs and other small livestock that perished in the floods. 27 ha of paddy fields

(a) (b)

Fig. 3. (a) Flooded School Playground and Temple Area, (b) Destroyed Irrigation Canals and Submerged Paddy Fields. *Source:* Courtesy of Mae Lana villagers.

(a) (b)

Fig. 4. (a) Weir 1 Got Clogged with Debris, (b) Weir 2 was Completely Destroyed. *Source*: P. Elstner.

were destroyed, accounting for 55% of the total paddy area in the Mae Lana Valley. Major elements of the irrigation infrastructure, such as canals and weirs, were partly damaged or completely destroyed (Figs. 4(a) and 4(b)).

Immediate Responses by Villagers and External Flood Relief Efforts

When the flood waters began to rise on the morning of 28 September, villagers organised cars to bring elderly villagers and children to higher ground in the upstream village Ja Bo. After the flood, villagers from Mae Lana and neighbouring villages helped together to clean up the village. Residents who had built their houses uphill and were not affected by the flood provided food for villagers who had lost their houses and whose rice harvest had been destroyed.

The Red Cross in the provincial capital Mae Hong Son was one of the first external organisations to provide emergency assistance. They gave food and water to the people who lost their houses or household items. Several government agencies and organisations followed suit and supported affected villagers in Mae Lana. Among those were the Pang Ma Pha District Office, the Royal Thai Army, the Community Organisations Development Institute and the Department of Public Works, who helped fixing the irrigation infrastructure (Fig. 5).

The Disaster Relief unit of the Department of Provincial Administration under the Ministry of Interior allocated 30,000 Thai Baht (THB) (US$ 750) to those households that lost their entire house and up to 20,000 THB (US $500) to households that had suffered severe damage to their properties. For damaged paddy rice fields, the victims received 43,750 THB (US $1,094) per ha which they could use for hiring tractors to level the land as well as to remove sediments and stones from their fields.

In sum, the immediate response to the flood disaster by villagers reflected a strong sense of unity and mutual assistance pointing to a high degree of collective action. The community's resilience to the flood

Fig. 5. Restored Irrigation Canal in Mae Lana, Supported by Various Thai Government Agencies. *Source*: P. Elstner.

impact was additionally strengthened by quick support from external agencies, which helped restoring the 'hardware', that is, the physical infrastructure that had been damaged, without government agencies interfering with the 'software', that is, the community-based management of water resources.

Causal Explanations for the Floods

Most of the interviewed farmers stated that the frequency of floods has not increased in recent years, although the magnitude of the floods in 2005 was exceptionally high. They believed that the cause of the flood was 'natural', that is, unusually high amount of rainfall, and that they could not do anything to prevent such kind of events. None of the interviewed farmers blamed the existing communal water management system for the disastrous impact of the floods. Only few farmers related the floods to increased conversion of forest into agricultural land in the upper slopes of the watershed. Yet, the village headman did not agree with that point of view, claiming that in former times, villages had cleared even larger tracts of forestland without facing a higher incidence of flood events.

Long-term Adaptation and Mitigation Strategies by Villagers in Mae Lana

The catastrophic flood event did not have a major impact on the communal water management system. The village headman, the village committee, the *gae muang* and all water users worked together to repair the irrigation system with the support from the district. Since the Weir 2 had been completely destroyed, all paddy field owners in the north of the community's residential area received water only from Weir 3 after the flood. In the area of Weir 3, the responsibility of water management remained with the farmers' group without the assistance by a *gae muang*.

The district office provided a simple device to measure the rainfall. In the future, if there was more than 100 mm rainfall in a day, the village headman would use a siren to warn the people against an immediate flood risk. The district also helped the community to dredge the Mae Lana River, widen the riverbed and construct a riverbank protection from stones and soil to prevent future flooding (Figs. 6(a) and 6(b)). Villagers

Fig. 6. (a) Widened Bed of the Mae Lana River, (b) River Bank Protection. *Source*: P. Elstner.

along the riverbank agreed to give up small parts of their paddy fields for this common purpose. The village headman stated in an interview that he had already planned to widen the river bed some years ago. Yet, at that time the villagers had rejected this measure for fear of losing part of their paddy field area.

The Community Organisations Development Institute (*phor or chaw*) supported farmers on a long-term basis in terms of house reconstruction and repair and career promotion, allocating a budget of 400,000 THB (US $10,000) managed by the village committee of Mae Lana. Flood victims could receive loans from this fund with an interest rate of 1% per year and the due date for repayment after five years.

Twelve families whose houses had been destroyed or severely damaged by the flood were given the permission by the Royal Forest Department (RFD) to resettle in their upland fields on the slopes surrounding the Mae Lana valley. The village headman had filed a formal request on behalf of the affected villagers.

In 2006, after another flood that destroyed a large part of the rice crop, farmers became increasingly aware of the risks of paddy rice production in the Mae Lana valley. Since then, several farmers have cleared forestland in the upper slopes of the watershed to grow upland rice to ensure their individual households' food security in the case of future flood disasters (Fig. 7).

Yet these individual adaptation strategies are likely to exacerbate the impact of future flood events on downstream paddy rice cultivation and thus run counter to the collective mitigation strategies in the river valley.

Fig. 7. Upland Rice Cultivation in the Upper Slopes of the Mae Lana Watershed. *Source*: A. Neef.

RESPONSES TO REPEATED FLOOD EVENTS IN THE CHIENG KHOI CATCHMENT, NORTHWEST VIETNAM

Governance of Forests, Irrigation Water and Agricultural Land Prior to the Flood

The Chieng Khoi commune is relatively well-endowed with land and water resources, and every household has paddy and upland fields as well as fish ponds. Aquaculture plays an important role in the agricultural system of the village. The main water source for irrigation is the Chieng Khoi Reservoir (created by a concrete dam built in 1974) and several smaller sources. The two outflows of the dam are the river and the main irrigation canal which allow two paddy crops per year in the flat plateau area. The hills upstream from the reservoir are classified as use forest, protection forest or reforestation forest. Farmers are allowed to collect non-timber forest products and can use timber for house construction after obtaining permission from the village headman.

In the upland areas surrounding the reservoir to the east and west farmers grow maize – their most important cash crop – and cassava for animal feed, both of which are rather prone to erosion. The plateau area downstream from the reservoir is used for paddy rice cultivation in combination with a complex system of fishponds (Fig. 8).

Typically for hydrological societies in East and Southeast Asia, the communal water governance system of the Black Thai in Chieng Khoi has developed over centuries. Like in North Thailand, the community-based water system is locally called *muang-fai*. Traditionally, the social organisation underlying the *muang-fai* system revolved around the water manager, the *liep (nam) na*. Similar to the case of Thai *muang-fai* communities in North Thailand, the duties of water managers in the traditional Black Thai society included the calculation of the amount of water and its allocation to individual farmers; coordination of the initial construction of weirs and canals; coordination of system maintenance; collection of fees and the performance of religious rituals.

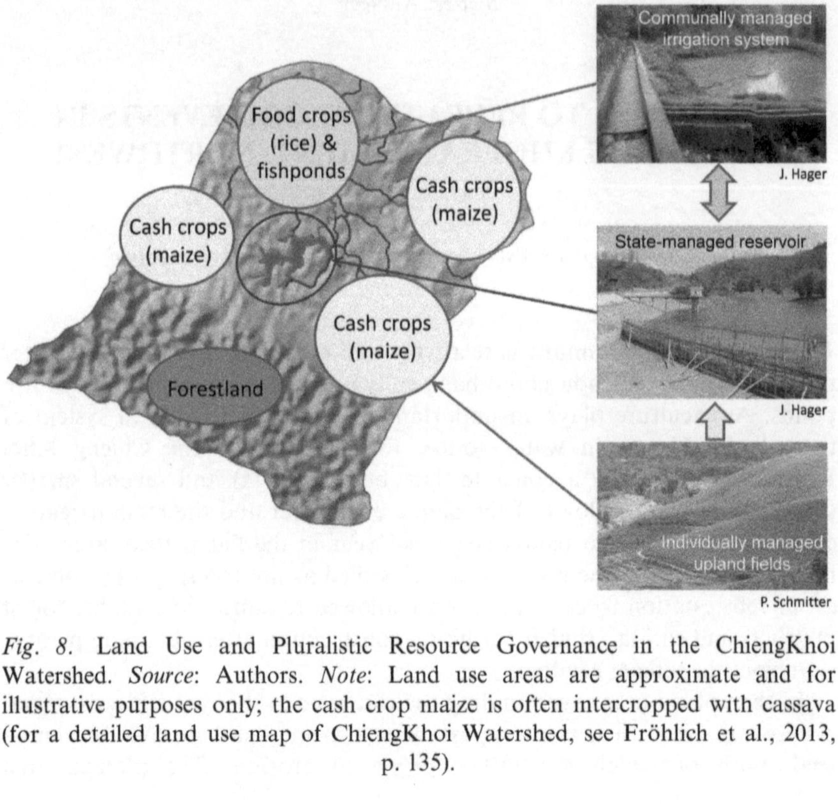

Fig. 8. Land Use and Pluralistic Resource Governance in the ChiengKhoi Watershed. *Source*: Authors. *Note*: Land use areas are approximate and for illustrative purposes only; the cash crop maize is often intercropped with cassava (for a detailed land use map of ChiengKhoi Watershed, see Fröhlich et al., 2013, p. 135).

The completion of a concrete dam in 1975 that formed the Chieng Khoi Reservoir marked a substantial change in local water governance. The farm households in Chieng Khoi did not have to provide any money, labour or in kind payment for the dam which was planned and built by the provincial government of Son La. Due to the high investment cost, the complexity of maintenance and the agricultural significance of the reservoir as a stable water resource, it is managed and maintained by employees of the provincial Department for Irrigation. The Provincial Department for Irrigation created the position of a reservoir manager under its direct guidance. His duties are to measure the water level on a daily basis, to regulate the water out-flow of the lake and to maintain and repair it. A contract was concluded between the user villages in Chieng Khoi commune and the Provincial Department of Irrigation regulating the use and management of water from the reservoir. It was determined that local users can influence the quantity of water flow through meetings between village heads and dam managers and also that users have to pay a fee for water. The contract stipulates an amount of 270.000 Vietnamese Dong (VND) (US$ 17) per ha for irrigation water for the spring rice crop and 190.000 VND (US$ 12) for summer crop irrigation.[3] Forty percentof this money is given to the reservoir manager and his team as remuneration and 60% is used to maintain the communal *muang-fai* system (Neef et al., 2006b).

The water management in Chieng Khoi Watershed is closely related to the administration at the commune, district and provincial level (Fig. 9). The village management board (*ban quan ly*) comprises the village headman, the treasurer and the accountant; the latter two serving as assistants to the headman. These three men are responsible for planning, implementing and controlling all common activities in the village. Their duties concerning water management are to regularly check the irrigation system, make the contract with the Chieng Khoi Reservoir manager and coordinate collective activities of villagers. However, all final decisions must be made in accordance with the cadres of the Communist Party – represented in the People's Committees at various administrative levels – as the most influential institutions. Besides the members of the management board on village level, the *Liep Na* (water manager) has the responsibility of maintaining minor weirs and canals, protecting fields from destruction by ruminants and observing the development of plant diseases within the village boundaries. The *Liep Na* is appointed by the headman for an undetermined time, with agreement of the villagers, and is obliged to inform the headman on a regular basis about the condition of the water and wet-rice

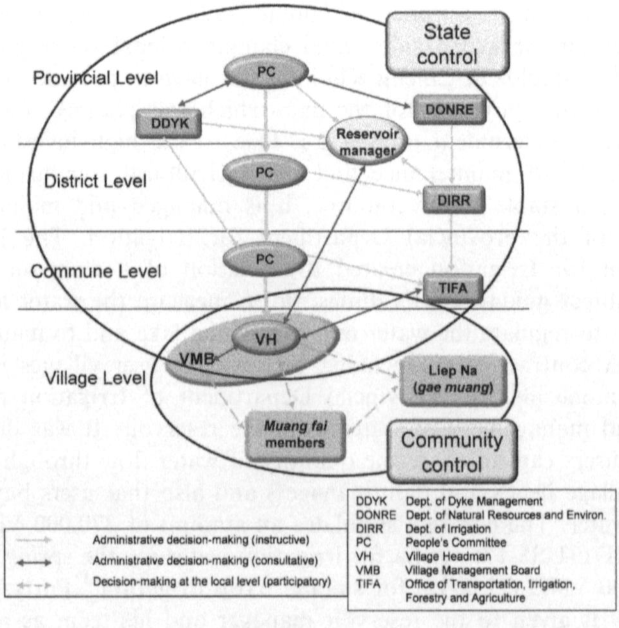

Fig. 9. Irrigation Water Management in the Chieng Khoi Watershed. *Source:* Authors' draft, partly based on Neef et al., 2006b.

systems. As compensation for his efforts he gets a fixed amount of rice after harvest (Neef et al., 2006b).

The responsibility for water allocation and collection of water fees is with the village headman who also decides on priorities in maintenance activities and serves as head of the respective village's 'disaster committee' (*ban phong chong lut bao*). However, the ultimate decision on how much water can run through the canals and be allocated to individual farmers rests with the commune chairman (Schad et al., 2012). The right to request bigger allocations or propose a different timing enables villagers to participate in water management decisions to a certain degree (*ibid.*). The duties of the lake manager and his team – who are not natives of the commune – are to measure the water level on a daily basis, to follow the weather forecast and to adjust the water management if necessary, to regulate the water outflow of the lake and to maintain and repair the dam infrastructure (Neef et al., 2006b).

A Disastrous Flood and Its Impact on Villagers' Livelihoods

In early October 2007, several days of continuous rainfall triggered by the off-seasonal typhoon Lekima caused a devastating flash-flood in the Chieng Khoi Watershed. Previous precipitation during the rainy season (May–September) had already overstretched the water-holding capacity of the soils, thus leading to excessive surface runoff into the reservoir. In a 36-hour period that started on 4 October, the rainfall amounted to around 165 mm (Schad et al., 2012). Twenty-four hours into this rain event, the spill-over of the reservoir had started to overflow, causing a dramatic swelling of the stream beneath the reservoir and turning much of the plateau area with its paddy rice fields into a disaster zone (Fig. 10).

For most villagers the flood came as a surprise, although interviewees who owned a TV stated they had learned about the forecast of exceptional rainfall in the news at least one day in advance (Schad et al., 2012). Most interviews in the study area confirmed that people underestimated the warnings, because of the unpredictability of weather patterns in this mountainous region of Vietnam and the unreliability of the local weather forecast. Respondents also stated that the weather in October is usually dry, which was confirmed by long-term average rainfall data.

Although the flood was described unanimously as the greatest flood disaster in living memory, only 17% of farmers among a random sample of 36 households in the Chieng Khoi commune reported significant individual losses to their properties or incomes.[4] Among these households, most had

Fig. 10. Rice Fields in ChiengKhoi (a) After Transplanting and (b) After the Flood in 2007. *Source*: K. Ogata (photo left); I. Schad (photo right).

Fig. 11. One of the Major Bridges in ChiengKhoi Watershed During the Flood. *Source*: I. Schad.

suffered from the loss of fish from their ponds or from sedimentation of their paddy fields, causing an average income loss of 3.3% among this group (Schad et al., 2012). The relatively low incidence of individual damages could be explained by the fact that maize – by far the most important cash crop – had already been harvested when the flood occurred (*ibid.*). Yet substantial damages to public infrastructure, such as bridges and roads, were reported, thus negatively affecting the communities as an entity rather than individual households (Fig. 11).

Immediate Responses to the Disastrous Flood Event

Emergency interventions in the Chieng Khoi Watershed were initiated right after the flood by the commune's People's Committee. The village headmen were asked to mobilise money, labour and reconstruction material from village resources to repair damages to roads, bridges, community houses, irrigation channels and other public infrastructure. Wood and bamboo were obtained from community forests, and individual households that held private forestland certificates had to contribute a certain amount of timber for reconstruction (Schad et al., 2012).

No support was offered by institutions beyond the commune level. Neither the Provincial Department of Dyke Management nor the National Disaster Fund was willing to provide assistance to individual flood victims or to the commune as a whole, although the village disaster committees had compiled detailed accounts of private and public damages and communal reports on the inflicted flood damages had been sent to higher administrative levels. The only tangible help provided for farmers was the remission of the irrigation water fee for the rice crop in the consecutive spring season 2008 (Schad et al., 2012).

Causal Explanations for the Floods

Local stakeholders attributed the flooding to general weather changes, that is, more frequent irregularities in the amount and timing of precipitation in the past 10 years. Commune officials stated that heavy rainfalls occurring after September would quasi automatically trigger a flood, as the reservoir reaches its storage capacity at the end of the rainy season; hence, excess water cannot be drained off in a controlled way. Few villagers acknowledged that the absorption capacity of upland soils has decreased due to deforestation and intensified upland cultivation with annual, erosion-prone crops, particularly maize. Most respondents in the interviews and – in an even more pronounced way – in their written essays attributed a large part of the blame on the reservoir manager. They stated that he had not followed the weather forecast, that he was too young for this job and that he had ignored the recommendations from the Department of Dyke Management to discharge water from the reservoir to increase its buffer capacity (Schad et al., 2012). The reservoir manager – on his part – claimed that deforestation and unsustainable upland farming practices – particularly the cultivation of maize on steep slopes without appropriate soil conservation measures – had induced sedimentation of the reservoir, thus reducing its buffer capacity. He also blamed the meandering of the river – caused by the establishment of fishponds along its way – as another factor for the unusual flood event (Schad et al., 2012).

Adaptation and Mitigation Strategies in the Aftermath of the Flood

The 'finger-pointing' behaviour that followed the flood disaster indicates that the major stakeholders in the Chieng Khoi Watershed were not ready to assume at least partial responsibility for the flood disaster. In connection with the lack of external help from state agencies at the district, provincial and national level, this apparent failure to share responsibility negatively affected the readiness of local stakeholders to engage in follow-up collective action. In addition, the diversification of risk – growing maize as a cash crop on erosion-prone slopes, while cultivating paddy rice and raising fish for home consumption in flood-prone plateau areas – appears to be a viable strategy at the individual households. Farmers who lost part of their rice yields or whose fish were flushed away from the ponds could still sell their maize harvest and ensure the food security of their families.

It is, therefore, not surprising that suggestions by the reservoir manager and researchers to plant more trees and other perennial crops around the reservoir were dismissed by the farmers in the group discussions. Hence, villagers in Chieng Khoi Watershed are likely to continue with a combination of adaptive risk-diversification strategies and reactive rehabilitation measures in the aftermath of disastrous flood events rather than adopting more proactive and anticipatory mitigation strategies.

SYNTHESIS AND CONCLUSION

The majority of the farmers in both study areas perceived floods as being caused by a set of exogenous factors that were beyond their individual and collective responsibility. Collective disaster-mitigation strategies were compromised or inhibited by individual households' adaptation, that is, risk-diversification strategies (in both countries) and by enhanced state control of water management (in the Vietnam case). Our findings suggest that local responses to flood disasters need to be understood as a complex interplay of individual livelihood strategies, collective responses and government interventions. Hence, the role of communal resource management and collective action in disaster response and recovery as well as in mitigation and adaptation can be ambivalent with hard-to-predict outcomes. We conclude that actors involved in flood-prevention strategies need to be aware of (i) local people's own causal explanations of flood events and (ii) the potential trade-offs between collective action towards flood mitigation, individual livelihood strategies and state intervention in water governance systems and flood response and recovery.

NOTES

1. For an interdisciplinary approach to investigating the flood disaster in Chieng Khoi, see Schad et al., 2012.

2. The *muang-fai* system is used on rivers, across which weirs (*fai*) are built up. The *fai* holds back water and directs it into a system of canals (*muang*) in which gates control flow rates. The way *fai* are built allows a certain volume of water to pass the barrier, while restricting the rate of flow sufficiently to raise the water level.
3. Fees for spring crop irrigation are higher because the irrigation period falls in the dry season when water levels in the reservoir are very low.
4. Data by Camille Saint-Macary; cited in Schad et al. (2012).

ACKNOWLEDGEMENTS

We are grateful to the villagers and all key informants in Mae Lana Watershed and Chieng Khoi Watershed for their hospitality and sharing of information. We would like to thank Pakakrong M. Williams and Pham Van Nghia for their support during the fieldwork in North Thailand and Northwest Vietnam. The study was conducted in the framework of the Thai Vietnamese German Collaborative Research Program 'Sustainable Land Use and Rural Development in Mountainous Regions of Southeast Asia' (The Uplands Program – SFB 564). The financial support of the German Research Foundation (Deutsche Forschungsgemeinschaft) is gratefully acknowledged.

REFERENCES

Brilly, M., & Polic, M. (2006). Public perception of flood risks, flood forecasting and mitigation. *Natural Hazards and Earth System Sciences*, *5*, 345–355.

Cohen, P. T., & Pearson, R. E. (1998). Communal irrigation, state, and capital in the Chiang Mai Valley (northern Thailand): Twentieth-century transformations. *Journal of Southeast Asian Studies*, *29*(1), 86–110.

Fröhlich, H. L., Ingwersen, J., Schmitter, P., Lamers, M., Hilger, T., & Schad, I. (2013). Water and matter flows in mountainous watersheds of Southeast Asia: Processes and implications for management. In H. L. Fröhlich, P. Schreinemachers, G. Clemens, & K. Stahr (Eds.), *Sustainable land use and rural development in Southeast Asia: Innovations and policies for mountainous areas* (pp. 109–148). New York, NY: Springer.

Heitz, C., Spaeter, S., Auzet, A.-V., & Glatron, S. (2009). Local stakeholders' perception of muddy flood risk and implications for management approaches: A case study in Alsace (France). *Land Use Policy*, *26*, 443–451.

Lebel, L., Manuta, J., & Garden, P. (2011). Institutional traps and vulnerability to changes in climate and flood regimes in Thailand. *Regional Environmental Change*, *11*(1), 45–58.

Lin, S., Shaw, D., & Ho, M. C. (2008). Why are flood and landslide victims less willing to take mitigation measures than the public? *Natural Hazards*, *44*, 305–314.

Neef, A., Chamsai, L., & Sangkapitux, C. (2006a). Water tenure in highland watersheds of northern Thailand: Managing legal pluralism and stakeholder complexity. In L. Lebel, X. Jianchu, & A. Contreras (Eds.), *Institutional dynamics and stasis: How crises alter the way common pool resources are perceived, used and governed* (pp. 64–88). Chiang Mai: Regional Center for Social Science and Sustainable Development (RCSD).

Neef, A., Elstner, P., & Hager, J. (2006b). Dynamics of water tenure and management among Thai groups in highland Southeast Asia: A comparative study of muang-fai systems in Thailand and Vietnam. *Proceedings of the Eleventh Biennial Global Conference of the International Association for the Study of Common Property (IASCP) "Survival of the Commons: Mounting Challenges & New Realities"*, Bali, Indonesia, 19–23 June 2006.

Ounvichit, T., Wattayu, S., & Satoh, M. (2008). Participatory management structure of large-scale people's irrigation system: The case of the Soprong *muang fai* system, northern Thailand. *Southeast Asian Studies*, *46*(1), 145–162.

Paripurno, E. (2006). Studies on cause and impact of flood disaster in Central Java, Indonesia: A community based disaster management perspective. *Science & Culture*, *72*(1), 32–39.

Schad, I., Schmitter, P., Saint-Macary, C., Neef, A., Lamers, M., La Nguyen, Hilger, T., & Hoffmann, V. (2012). Why do people not learn from flood disasters? Evidence from Vietnam's northwestern mountains. *Natural Hazards*, *62*(2), 221–241.

Sirivongs Na Ayudhaya, A. (1979). *A comparative study of traditional irrigation systems in two communities of Northern Thailand.* PhD thesis, Social Research Institute, Chulalongkorn University, Bangkok.

Surarerks, V. (1986). *Historical development and management of irrigation system in Northern Thailand.* Chiang Mai: Faculty of Social Sciences, Chiang Mai University.

Surarerks, V. (1998). The müang-faai-irrigation system in northern Thailand: Historical development and management. *Tai Culture*, *3*(2), 37–48.

Tang, S. Y. (1992). *Institutions and collective actions: Self-governance in irrigation.* San Francisco, CA: Institute for Contemporary Studies Press.

Tran, P., Marincioni, F., & Shaw, R. (2008a). Catastrophic flood and forest cover change in the Huong river basin, central Vietnam: A gap between common perceptions and facts. *Journal of Environmental Management*, *91*, 2186–2200.

Tran, P., Marincioni, F., Shaw, R., Sarti, M., & An, L. V. (2008b). Flood risk management in Central Viet Nam: Challenges and potentials. *Natural Hazards*, *46*(1), 119–138.

Tan-Kim-Yong, U. (1995). *Muang-fai communities are for people: Institutional strength and potentials.* PhD thesis. Bangkok: Social Research Institute, Chulalongkorn University.

Wyatt, D. K. (1984). *Thailand: A short history.* Chiang Mai: Silkworm Books.

CHAPTER 6

FARMERS' RESPONSE TO DROUGHT IN NORTHWESTERN BANGLADESH

Umma Habiba and Rajib Shaw

ABSTRACT

Of all the natural disasters, drought is the most gradual and the most hard to predict. However, this insidious disaster continually affects the lives and livelihoods of farmers living in drought-affected areas. The northwestern part of Bangladesh is recognized as being more severely affected by drought than the rest of the country, as drought is a recurring event in this area. It has substantial impacts on agriculture and causes great suffering for farmers – in particular, poor and small farmers, who are more vulnerable to drought. Therefore, this study tries to illustrate farmers' existing coping practices with regard to drought. It also addresses their prioritized adaptation practices, which are based on local context and available resources. This study not only focuses on the implementation of these adaptation practices from the national to the local level, but it also mentions various roles of stakeholders and a definite timeframe for each adaptation practice.

Keywords: Drought; farmers; responses; adaptation practices; northwestern Bangladesh

Risks and Conflicts: Local Responses to Natural Disasters
Community, Environment and Disaster Risk Management, Volume 14, 131–161

ISSN: 2040-7262/doi:10.1108/S2040-7262(2013)0000014012

INTRODUCTION

According to the National Drought Mitigation Center (NDMC) (2006), drought is an insidious onset disaster that takes place in Bangladesh more frequently because of climate change. Habiba et al. (2012) further reported that not only climate change but also several human activities foster drought in Bangladesh. It is a recurrent event in some parts of the country, but the northwest region is more prone to drought area because of its high rainfall variability (Shahid & Behrawan, 2008) – the area is relatively dry, receiving much lower rainfall amounts as compared with the rest of the country (Paul, 1998). Therefore, drought occurs in this region more frequently. It is being reported more often in the Rajshahi, Chapai Nawabganj, Naogaon, Rangpur, Bogura, Pabna, Dinajpur, and Kustia regions because of their moisture-retention capacity and infiltration-rate characteristics.

Climatic variability and the nonavailability of surface water resources make drought a more common occurrence in the northwestern part of Bangladesh (Habiba, Shaw, & Takeuchi, 2011a, 2011b). Also, high rainfall variability and high temperatures leave this region prone to drought. Every year this region experiences a dry season for seven months, from November to May, when rainfall is normally low. Hence, drought mostly occurs in the pre-monsoon (March–May) and post-monsoon (October–November) seasons in Bangladesh (Banglapedia, 2006). Moreover, ADB (2005) highlighted that with climate change, lower precipitation during the dry season has the potential to increase the drought risk in Bangladesh. But the type and extent of drought varies according to the quantity and type of precipitation.

During the past 50 years, Bangladesh has suffered about 20 periods of drought. Major droughts occurred in 1966, 1969, 1973, 1978, 1979, 1981, 1982, 1989, 1992, 1994, 1995, 1998, 2000, 2006 and 2009. However, the consecutive droughts of 1978 and 1979 directly affected 42% of cultivated land and reduced rice production by an estimated 2 million tons. The losses due to drought in 1982 were more than double the losses caused by the floods that occurred in the same year. The drought in 1997 caused a reduction of around 1 million tons of food grain of which 0.6 million tons were transplanted aman. Furthermore, in 2001, Iqbal and Ali reported that every year 0.45 million ha of land are affected by very severe drought during the rabi season, while 0.40 million ha and 0.34 million ha of land are very severely affected during the pre-kharif (March 16–June 30) and kharif (July 1–October 15) seasons, respectively. On the other hand, Tanner et al. (2007) reported that about 2.7 million ha of land in Bangladesh are vulnerable to annual drought. The Climate Change Cell of MoEF (2009) reported that about 83% of 12.49 million ha of T-aman cultivable land and 9.32 million ha of rabi crop land are affected by drought during the pre-kharif season and rabi season (October 16–March 15), respectively.

Aside from the impact to agriculture, 5.038 million people are threatened by the effects of drought. Farming communities, in particular, suffer from the effects of drought, as they are the key stakeholders for food crops in Bangladesh. Throughout the country, about 2 million small farmers and 2.4 million rural wage laborers are vulnerable to the effects of very severe to severe kharif droughts. For example, drought increases the risk to food security, increases the risk of illness, reduces the sources of drinking water, causes increases in regional migration, and causes loss of livestock. In extreme cases, it increases the chances of seasonal food crises such as monga (famine). To cope with drought, farmers in this region have carried out a number of local as well as institutional responses to drought. Sadly, facilities and activities to cope with drought are very limited according to the demands placed on them. The successful implementation of drought risk management policy necessitates greater involvement and support from groups such as governmental and nongovernmental, as well as private, organizations. At the same time, involvement of the rural community in drought risk management planning and implementation programs is important for the formulation of policy in this regard.

Therefore, this chapter highlights the causes, types, and impacts of drought in Bangladesh. We provide a case study that reveals the current, ongoing responses of farmers in the northwestern region. The chapter also discusses the prioritized drought adaptation practices (DAP) of farmers to cope better with drought in the future and how these DAP should be linked and translated at national and local levels in a sustainable way.

CAUSES OF DROUGHT

- The Northwestern region is composed of uplifted terraces of Pleistocene sediments called "Barind tracts."
- The Northwestern region is hemisphere dominated by summer weather. So, this region is generally warm and humid. There is a very little rainfall and the weather remains hot during the day but becomes cooler late at night.
- There is little rainfall in this region as compared with the rest of Bangladesh, with annual average rainfall varying between 1,400 and 1,900 mm; 13% of rainfalls occur during the dry season.

- This region of Bangladesh also has higher temperatures than the rest of the country. The mean temperature ranges from 25°C in the hottest season to 9–15°C in the coolest season. In the summer season, the northwestern region experiences temperatures as high as 45°C or even more.
- Evapotranspiration in this region is more than 0.5 times higher than the rest of the country during dry season.
- During the dry season, the water deficit is too high to meet the demands of the crops.
- Less vegetation having lower soil moisture content in the area.
- The northwestern region is bounded by the jamuna (the Brahmaputra River to east and the Ganges River to the south). The major rivers of the northwest region are the Teesta, Upper Karatoya, Atrai, Karatoya, and Bangali. Pond, beels (A beel is a term for a pond with static water.), and canals are the larger water bodies and are generally connected to the rivers during monsoon season. Most of the rivers of this region flow from very steep to flat ground. Because of the consequences of climate change as well as of drought, the water bodies in these areas are going dry, river beds are filled with sand, and water flow in the rivers is decreasing. Therefore, the surface water is almost empty everywhere during dry season. Ultimately, it heightens the changes of droughts occurring in this region.

TYPES OF DROUGHT

Depending on the occurrence, there are four types of drought in Bangladesh. The character of each drought is described in Table 1.

DROUGHT IMPACTS

According to Wilhite, Hayes, and Knutson (2000) and Coleen, Laing, and Monnik (2006), drought impacts are diverse and can be classified broadly as economic, environmental, and social. But in the case of Bangladesh, drought affects the northwestern farmers tremendously because they depend on agriculture for their subsistence. Therefore, agriculture as well as rural people's lives are immensely affected by the results of droughts.

Table 1. Characteristics of Drought in Rajshahi District of Bangladesh, Where BMDA Activity is Currently Ongoing.

Type of Drought	Characteristics
Meteorological drought	Meteorological drought is the level of dryness and the duration of the dry period. Atmospheric conditions that result in deficiencies of precipitation change from area to area.
Hydrological drought	Hydrological drought is associated with the effects of periods of precipitation shortages on the water supply. Water in hydrological storage systems such as reservoirs and rivers are often used for multiple purposes, such as flood control, irrigation, recreation, navigation, hydroelectric power, and wildlife habitats. Competition for water in these storage systems escalates during a drought, and conflicts between water users increase significantly.
Agricultural drought	Agricultural drought mainly affects food production and farming. Agricultural drought and precipitation shortages lead to soil water deficits, reduced groundwater or reservoir levels, and so on. Deficient topsoil moisture during the planting season may stop germination, leading to low plant populations, which causes low to no yield.
Socioeconomic drought	Socioeconomic drought occurs when the demand for an economic good exceeds the supply as a result of a weather-related shortfall in the water supply. The supply of many economic goods, such as water, forage, food grains, fish, and hydroelectric power, depends on weather. Because of the variability of the climate, the water supply is sufficient in some years but not satisfactory to meet human and environmental needs in other years. The demand for economic goods is increasing as a result of increasing population. Supply may also increase because of improved production efficiency and technology.

Source: Habiba (2012).

Agriculture

Drought affects various crops during three different cropping seasons. A wide range of agricultural crops are grown in the rabi and pre-kharif periods and become vulnerable to drought to varying degrees. Rahman and Biswas (1995) claimed that crop yield in drought-affected areas of northwest Bangladesh was reduced by up to 60–70%. T-aman alone is hampered by various degrees of drought during the rabi season and pre-kharif season, and drought affects 7.15 million ha of land, which is 91% of the total T-aman land. Nasreen and Hossain (2002) state that drought affects standing crops and water supplies and impedes plant growth, leading to loss of productions, food shortages, and famine. And drought affects not only the rice crops but also other agricultural crops, such as jute, wheat, corn, potatoes, sugarcane, and different types of pulses and oilseeds and vegetables and causes significant damage to production in areas where irrigation is limited (Ahmed, 2006).

Fisheries

The frequency of extreme droughts has a disproportionate effect on fish habitats and populations, and the incidence of diseases among fish is expected to rise. Drought causes surface water bodies – such as canals, ponds, beels, and rivers – to dry up. It also hinders fishing activities and causes difficulties for those who depend on fishing for their livelihood. Drought hampers hatchery operations, fish production, and the livelihoods of a wide range of people directly and indirectly involved with fisheries and aquaculture.

Livestock

Drought causes an unfavorable environment for livestock. It also leads to a scarcity of food for domestic livestock. During periods of drought milk production, reproductive health, even the growth of livestock is adversely affected by drought. In severe cases, black quarter and anthrax-type diseases appear in drought-affected areas, leading to loss for those who are dependent on livestock for their livelihood.

Social Life

Drought also has a tremendous impact on social lives. During periods of drought, food consumption and food-purchasing capacity are reduced because of insufficient food production and the increases in food prices. Safe drinking water is scarce during periods of drought. And women face increased hardship because they must fetch drinking water from further away than usual. Children lose interest in going to school, resulting in an increase in the dropout rate.[1] And with no work available in the drought-prone area, seasonal migration occurs.

Health

Drought causes major health problems for people living in drought-prone areas. In the absence of drinking water, people are forced to use pond water, which can lead to various types of waterborne diseases such as dysentery, diarrhea, and hypertension. Even in conditions of high temperature and humidity, dehydration is a problem that especially affects the elderly and children. Drought also leads to an increase in chronic energy deficiency among members of the agricultural labor force because of the decrease in food consumption.

Environment

Drought has substantial effects on ground aquifers. The dependency of farmers on groundwater for irrigation purposes has increased in the past decade. This excessive use of groundwater for irrigation purposes coupled with domestic uses cause depletion of the groundwater level during the dry season. This in turn causes a great threat to the irrigated agricultural system because of the drawdown of the groundwater level leading to environmental problems such as heavy metal contamination and salinity. It is stated that the ground aquifer level has dropped by 8.95–to 18.56 m in some regions of northwestern Bangladesh (MoEF, 2002).

Drought also acts as a catalyst of land degradation by reducing soil moisture and water-retention capacity. It dries out topsoil leading to an effective loss of soil structure and aggregation. Moreover, drought depletes organic soil contents and reduces microbial activity and overexploitation of sparse vegetation. Drought also causes loss of biodiversity by hampering the microbial activities, causes the extinction of some species, damages plants and animals, and increases the incidence of diseases by bringing pathogen and parasites.

FARMERS' RESPONSE TOWARDS DROUGHT: A CASE STUDY

Study Location

The study area covers two districts severely affected by drought – Rajshahi and Chapai Nawabganj, located in the northwestern part of Bangladesh. It comprises 14 upazilas (subdistricts) in these two districts (9 upazilas in the Rajshahi district and 5 in the Chapai Nawabganj district) (Fig. 1). With an area of 4151.31 km^2, this region supports a population of 3.47 million, of which 52% are literate; 77% of the population is primarily depending on agriculture. The topography of this area is mainly flat with an average elevation of 25 m above mean sea level.

The average annual total rainfall in this area is 1,329 mm. According to Habiba, Takeuchi, and Shaw (2010), the annual total rainfall of drought-prone areas showed a distinct variation as compared with the annual total rainfall of Bangladesh as a whole – the difference is almost 1,000 mm. Almost 87% of rainfall occurs during the monsoon season (June–October) and 13% of rainfall occurs during the dry season (November–May). Therefore, evapotranspiration in this region is more than 0.5 times higher during the seven-month-long dry season than during the monsoon season, resulting in the demand for crop water during the dry months of the year being higher than the total rainfall. And comparatively high temperatures prevail in this area – between 25 and 35°C. Some of the hottest days get as high as 45°C or more.

The households of this area depend on groundwater for both domestic and irrigation purposes. Almost 75% of the irrigation water comes from groundwater and the rest comes from surface water (Bari & Anwar, 2000).

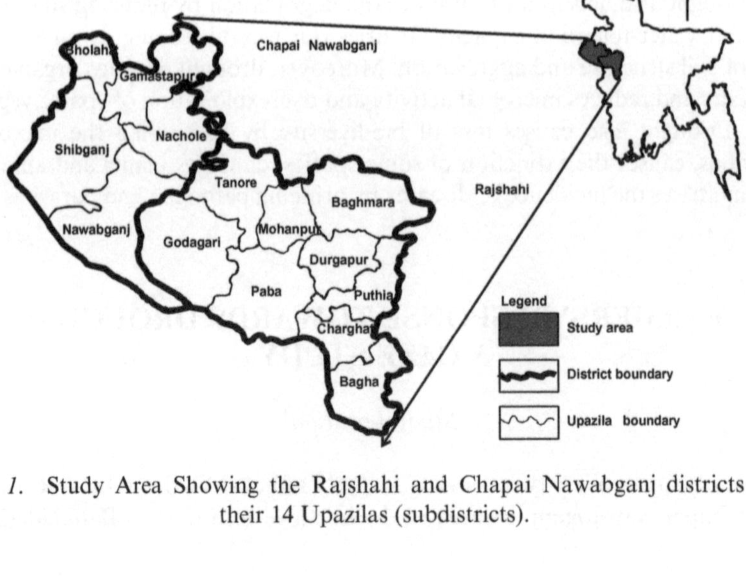

Fig. 1. Study Area Showing the Rajshahi and Chapai Nawabganj districts and their 14 Upazilas (subdistricts).

According to Shahid and Hazarika (2009), about 59% of cultivable land in this region is under irrigation. Since all the canals and rivers dry up during the dry season, groundwater is the only source of irrigation water in this region, and people become absolutely dependent on this source (Shahid & Hazarika, 2009). As a consequence of using excessive groundwater for both purposes, depletion of groundwater storage occurs during the dry season in northwestern Bangladesh (Habiba et al., 2010). This causes a great threat to the irrigated agricultural system because of overdrawn aquifers, lowering of water tables, and reduced stream flow.

Study Method

A primary field visit was done before conducting the questionnaire survey of the farmers. After visiting the area and consulting with the Upazila Agricultural Officer (UAO) of each upazila, the study areas were categorized as irrigated and nonirrigated (Table 1). On the basis of the irrigation facilities recommended by the UAO, the irrigated areas are those where irrigation is mainly operated by Barind Multipurpose Development Authority (BMDA) through deep tube wells. (It should be noted that BMDA is the governmental agency that only works in the northern region of Bangladesh for supplying irrigation as well as drinking water.) On the other hand, nonirrigated areas are those where BMDA activities are absent and farmers must devise their own ways to irrigate their fields. Fig. 2 shows one of the irrigated villages.

A total of 718 questionnaires were randomly distributed to and collected from farmers from both irrigated and nonirrigated villages of all upazilas in the two districts. The questionnaire focused mainly on the existing DAP of farmer, supports from institution during periods of drought period, and prioritized DAP of farmers that will help them to cope with drought in the future. In addition, two workshops were conducted at the national and local levels to identify the timeframe and the stakeholder's role for each DAP (Table 2).

Results and Discussion

Farmers Socioeconomic Characteristics

Socioeconomic characteristics such as gender, age, education, income level, and landholding size lead individuals to take certain adaptive measures

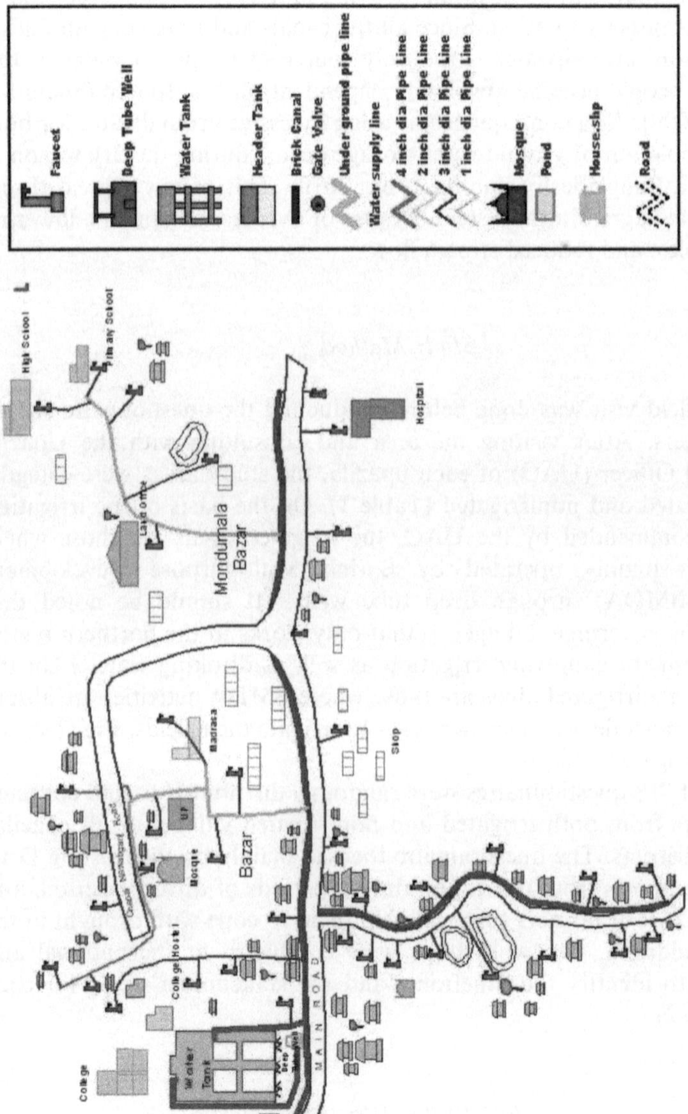

Fig. 2. Map Showing BMDA Coverage in the Sadipur Village of the Tanore Upazila in the Rajshahi District. *Source:* Barind Multipurpose Development Authority (2010).

Table 2. Irrigated and Nonirrigated Villages in the Study Area.

District	Upazila	Union	Irrigated Village	Nonirrigated Village
Rajshahi	Paba	Nouhata	Baroipara	Teghore
	Mohanpur	Dhoroil	Mollakuri	Palsa
	Godagari	Matikata	Vatapara	Kadipur
	Charghat	Municipalities	Mukterpur Dofadorpara	Raotha
	Bagha	Monigram	Hassenpur	Vanukar
	Puthia	Shilbaria	Malipara	Dhononjoypara
	Baghmara	Gobindopur	Khejur	Maria
	Tanore	Mundomala municipalities	Shadipur	Dhaudhum
	Durgapur	Nowpara	Palsa	Durgadhaho
Chapai Nawabganj	Nawabganj	Jhilim	Bilbaitha	Boro pukuria
	Shibgan	Durlovpur	Notungram	Baro rosia
	Gomastapur	Radhanagar	Khokadanga	Borodadpur
	Nachole	Fotapur	Alisapur	Takahara
	Bholahat	Gohalbari	Gohalbari	Borogachi

during the crisis period. Survey findings show that most of the farmers in both areas are male heads of household who engaged in farming activities and that they are in the middle aged (from 26 to 50 years) group, indicating that they are still in the active age bracket. Middle-aged farmers have been found to be more knowledgeable about best practices and may be more willing to bear risk and adapt to better farming techniques because of their longer planning horizons. It has also been observed that in both areas more than half of the farmers have primary or secondary educations. Very few farmers in the two areas have completed college or graduate degrees. Regarding income level, 28% and 31% of farmers have incomes between Tk 21,000 and 40,000 (1 USD = 80 Tk) in irrigated and nonirrigated areas, respectively. Only 18% of farmers in the two villages have incomes higher than Tk 80,000. In terms of landholding size, farmers in the two villages belong to the small, marginal, and landless category, which leaves them unable to make the decisions about adaptation practices during periods of drought. Even in cases of farmers with tenure status, the percentage of tenant farmers is higher in nonirrigated villages than in irrigated villages. This also means that owner farmers in irrigated villages have more capacity to use adaptation practices in their fields than in nonirrigated villages because they have their own land.

Drought Impacts on Farmers' Livelihood

The majority of farmers (more than 90%) from both irrigated and nonirrigated areas stated that their main sources of income and food are affected by drought (Fig. 3). Drought has an especially bad effect on high-yielding rice varieties such as boro. For instance, during dry season, boro crops require optimal watering for their flowering and maturity stage. But farmers from both areas could not provide sufficient water because of a disturbance in the electric supply, leading to crop failure and farmers not getting their expected crop yield.

Apart from the agricultural losses, more than half of the farms with irrigated and nonirrigated crops noted that drought has a tremendous impact on their daily life. During periods of drought, farmers incurred debt to save their fields. And sometimes these farmers became involved in water-related conflicts because of the unavailability of water during the dry season (Habiba et al., 2011b). Hence, drought increases the conflict between different groups of water users – those who need just drinking water and those who also need water for irrigation. In addition, drought causes various health problems. For example, disorders such as fever, headache, dysentery, diarrhea, and hypertension occurred during the dry season. Moreover, in the absence of drinking water, women and girls of farming families spent more time fetching safe drinking water from farther distances. In some cases, families drank pond water, which can accelerate various types of waterborne diseases.

In terms of animal husbandry, drought aggravates the availability of all types of livestock feed, including rice straw and green grass, leading to a lack of required nutrients. This will ultimately affect milk and meat

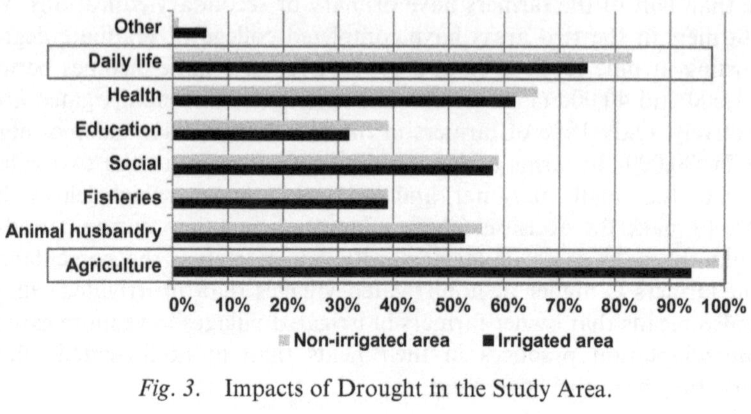

Fig. 3. Impacts of Drought in the Study Area.

production. The poor level of nutrition reduces the general immunity of the livestock. Therefore, cattle become more susceptible to various diseases. And in severe cases, livestock, cattle, and poultry have suffered from heat stroke.

Existing Drought Adaptation Practices of Farmers

In the study area, the majority of farmers (more than 90%) from irrigated and nonirrigated villages practiced different types of agronomic management to cope with recurrent droughts (Fig. 4). They perform primarily agronomic practices that help them to improve agricultural productivity. The common agronomic practices observed in the study area are manuring and composting, seedbed method, ail lifting, tillage, and shedding. These practices help to improve soil quality, add organic matter to the soil, hold rainwater during the rainy season, and distribute rainwater uniformly into the fields.

It is apparent from Fig. 4 that farmers are practicing water harvesting through re-excavation of ponds, kharis (a khari is lengthy pond-like structure with a length of 2–3 km and a width varying from 10 to 15 m), or canals. Farmers of nonirrigated areas have less scope to use harvested water because of the absence of BMDA's irrigation facilities in their area.

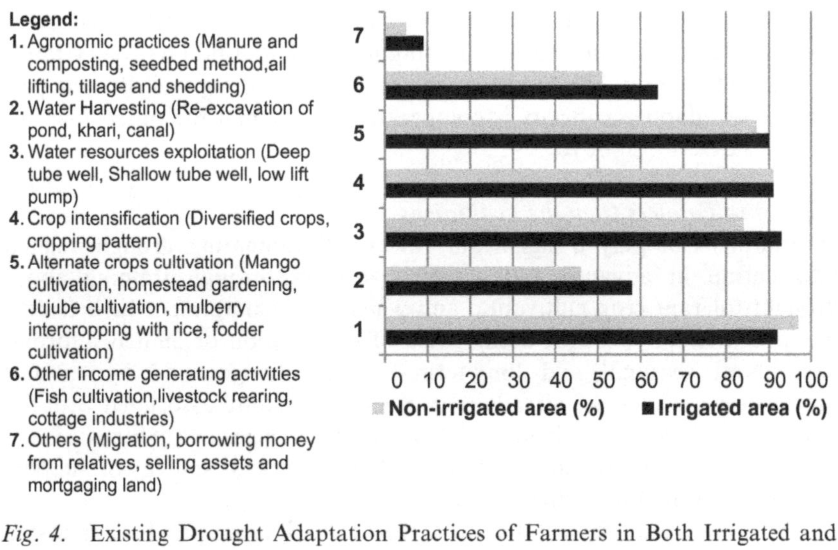

Fig. 4. Existing Drought Adaptation Practices of Farmers in Both Irrigated and Nonirrigated Areas Within the Study Area.

Apart from the water harvesting, farmers utilized various type of exploitation of water resources, such as deep tube wells (DTW), shallow tube wells (STW), and low lift pumps (LLP). Fig. 4 also shows that the majority of farmers (more than 90%) in irrigated areas used water from DTW, whereas, in nonirrigated areas, more than 80% of all types of farmers were supplying water through STW and LLP. However, the percentage of use of these methods is lower than the other adaptive practices currently farmers perform.

A majority of farmers (more than 90%) from both areas said that they diversified their crops to cope with drought, growing plants such as sugarcane, different types of pulse and oil crops, vegetables, and different fruit crops, such as mango and jujube. In addition, farmers in both areas followed different cropping patterns in their fields. For example, planting boro–aus–rabi crop (mustard–potato–onion) is the most popular practice in irrigated areas, whereas planting potato–aus–aman/vegetables is the practice in nonirrigated areas. Likewise, farmers from both areas followed mixed cropping or intercropping in their fields because of the opportunity to grow more than one crop in the same field. About 80% of farmers in the two areas cultivate various alternatives to rice crops. They set up mango orchards, jujube orchards, and homestead gardening, and plant mulberry with rice, etc. in their fields.

Some farmers in both areas utilized other income-generating activities, such as business, wage labor, services, and construction works, and other nonagricultural farming, such as rearing livestock and poultry, dairy farming, cattle fattening, fish culturing, and making cow dung fuel. But the percentage of adaptive capacity is comparatively less than the other adaptive practices described above.

Support to Farmers from the Institutions

Institutions can play a significant role in disseminating drought-related information in advance and assisting farmers through training about drought-tolerant crop cultivation and watershed management before, during, and after the drought event. Lack of information or limited information about seasonal and long-term climate changes and agricultural production can increase already high risks from failure associated with the uptake of new technologies and adaptation measures (Jones, 2003; Kandlinkar & Risbey, 2000). Here we depict farmers' sources of support and information during drought period.

The survey results reveal that more than 75% of farmers in both irrigated and nonirrigated areas acquire most of their weather-related information

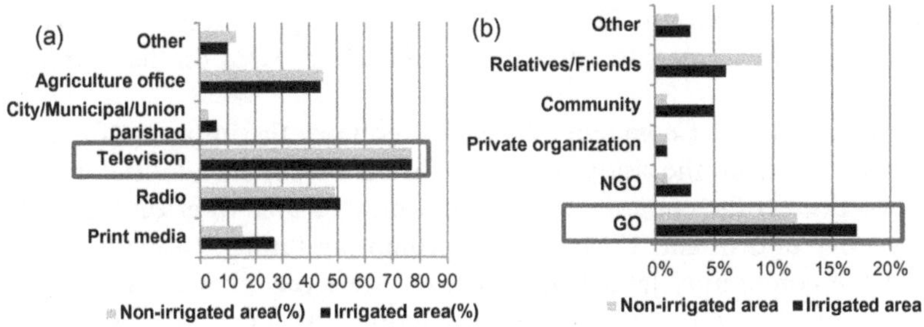

Fig. 5. Farmers' Drought-Related Information (a) and Sources of Support (b) During Periods of Drought.

from television (Fig. 5(a)). Television and radio stations receive early warning information about heavy rainfall, droughts, and tropical cyclones and associated storm surges from Bangladesh Meteorological Department (BMD) and Flood Forecasting and Warning Center (FFWC) and broadcast this information throughout the country. Farmers get less drought-related information from the agriculture office and the print media.

The survey questionnaire asked farmers about their sources of support during drought periods. The results in Fig. 5(b) shows that 17 and 12% of farmers in irrigated and nonirrigated areas, respectively, mention that they mainly receive support from governmental offices and 6 and 9% from relatives and friends. But, the sources of support are very limited in the study area because of the very few governmental as well as other organizational supports during periods of drought.

Prioritized Drought Adaptation Practices of Farmers

According to Rojas Blanco (2006), communities have the right to be informed about the ramifications of climate change and they state that the farmers in these communities are capable of generating solutions likely to work best in their localities. In this regard, the community is considered as the center of disaster risk reduction (DRR), and community actions make disaster reduction most effective. Therefore, community participation and control is essential for any successful implementation, orientation, and maintenance of any DRR project (Habiba & Shaw, 2012). In this section, we attempt to figure out farmers' prioritized DAP. Depending on the local context, time orientation, and available resources, farmers identified the top DAP, which we call 1st priority practice (P-1). Likewise, 2nd (P-2), 3rd (P-3), and 4th (P-4) priority practices indicate important, less important, and much less important DAP. Multiple-choice-type adaptive options were provided for the farmers to select the suitable DAP based on their priorities.

Prioritized DAP at Community Level. To counter the impacts of drought, survey results from the study show that farmers recommend different adaptive practices at the community level that may be able to effectively reduce the severity of drought. Furthermore, promoting farmers' adaptive capacities at the community level helps them to effectively alleviate the negative impacts of drought. Therefore, considering the farmers' responses to drought, the adaptive practices are listed and prioritized according to their demands (Fig. 6 and Table 3).

Among 15 drought adaptation options, 50% of farmers in both irrigated and nonirrigated areas agree with 13 DAP as the topmost or 1st priority practice (Fig. 6). They differ only in two DAP (i.e., taking preventive measures and aid from institutions). In terms of taking preventive measures, 47% of farmers in irrigated areas prioritize dairy farming, which would be a suitable adaptive measure for them to cope with drought. On the other hand, 72% of farmers in nonirrigated areas recommend community savings and credit systems to help them start a business or other income-generating activity. Moreover, it is interesting that 80% of farmers in the two areas approve and emphasize only three DAP that are highly suitable at community level. These are presented in Table 4 and described as follows:

Establishing a mango orchard is considered a suitable adaptive practice at the community level. The reasons for choosing this practice in both villages are the favorable weather and soil conditions for mango cultivation. In addition, farmers are getting more interested in cultivating different varieties of mango because of high profit with low investment, as reflected in the huge demand by affluent farmers and traders for taking leases on orchards. Therefore, at the community level, farmers in this region are interested in cultivating mangoes as a main cash crop instead of other crops. Habiba et al. (2011b) found that the cultivated land in this area is now rotating into fruit crop fields such as mango orchards. Moreover, the mango-cultivation area has increased from 4,650 ha in 2001 to 8,667 ha in 2011.

Community health care service is important as compared with a primary health care facility because it focuses on basic services to improve health status, which builds community resilience and provides the foundation for responding to emergencies. It reduces vulnerability, protects health facilities and services, and scales up the response to meet the wide range

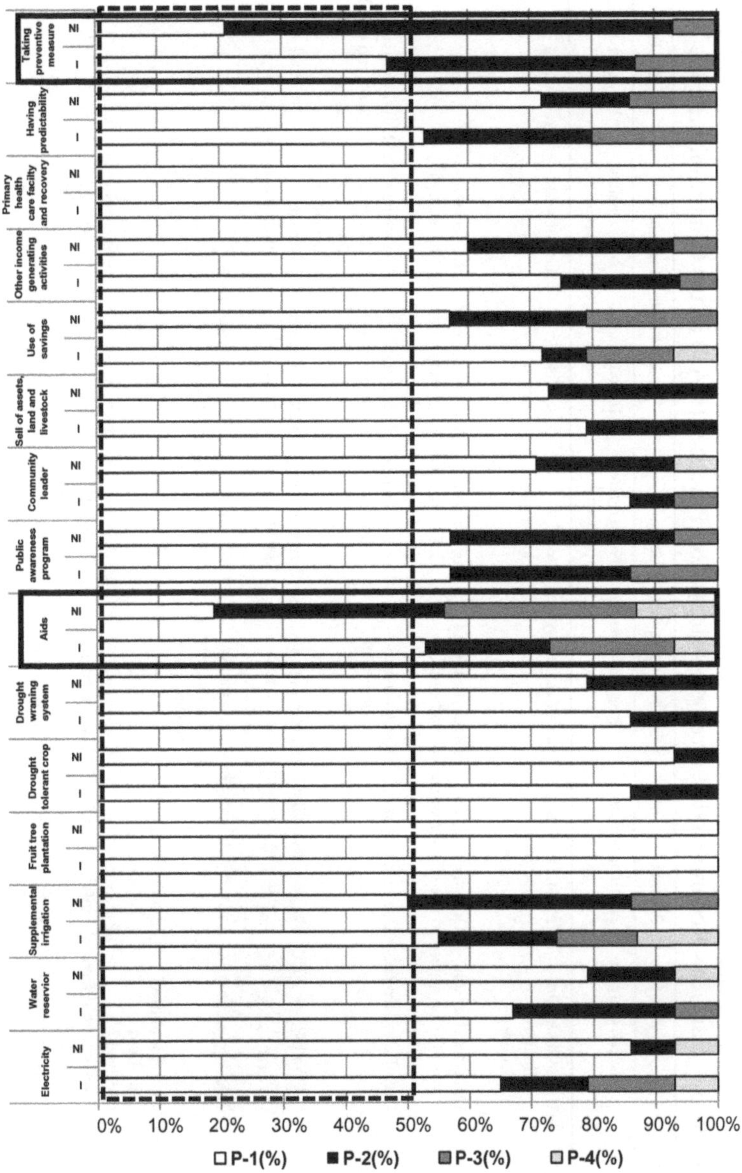

Fig. 6. Recommended and Prioritized DAP of Farmers at the Community Level.

Table 3. Farmer's Recommended Drought Adaptation Practices (DAP) at the Community Level.

Serial No.	Name of the Drought Adaptation Option	1st Priority Practice		2nd Priority Practice		3rd Priority Practice		4th Priority Practice	
		Irrigated area	Nonirrigated area	Irrigated area	Nonirrigated area	Irrigated area	Nonirrigated area	Irrigated area	Nonirrigated area
1	Alternative energy sources for irrigation (electricity)	Use of diesel for irrigation purposes	Use of diesel for irrigation purposes	Use of electricity	Use of electricity	Use of solar pump	Use of solar pump	Fallow land	Fallow land
2	Water reservoir	Drag the river and use of river water	Drag the river and use of river water	Digging of Pond and canal	Dig the pond, khari and use pond water	Construction of groundwater dam	Rain water harvest	–	–
3	Supplemental irrigation	Use of plastic pipe	Use of plastic pipe	Use stored sub-surface water	Use sub-surface stored water	Irrigation through STW	Use shallow tube well for irrigation	Use re-excavated pond and canal's water	–
4	Taking preventive measure	Dairy farm	Community savings and credit system	Community savings and credit system	Establish dairy farm	Establish grain bank	Establish grain bank	–	–
5	Drought-tolerant crops	Vegetable gardening	Vegetable gardening	Cultivation of banana, guava, litchi	Cultivate cassava crop	Cultivate cassava plant	–	–	–
6	Fruit tree plantation	Establish mango orchard	Establish mango orchard	–	–	–	–	–	–
7	Having predictability	Received information from TV, radio	Received information from TV, radio	Exchange drought knowledge	Exchange drought knowledge	Newspaper	From print media	–	–
8	Drought warning system	Establish drought information center	Establish drought information center	Early warning system	Public advisory committee	–	–	–	–

9	Community leader/ Imam	Extension worker	Extension worker	Community leader	Community leader	Imam	NGO worker	NGO worker	–
10	Primary health care facility	Community health care service	Community health care service	–	–	–	–	–	–
11	Other income generating activities	Establish dairy farm	Establish dairy farm	Community business, nursery	Business	Batik house	Batik house	–	–
12	Public awareness program	Public awareness program	Public awareness program	Drama regarding drought	Drama regarding drought	Motivational tour	Motivational tour	–	–
13	Aids	Cash	Electricity	Electricity	Agricultural equipments	Agricultural equipments	Cash	Livestock	Livestock
14	Use of savings	Save money	Save money	Save rice	Save quality seeds	Rearing livestock/ poultry	Rearing livestock	Handicrafts	–
15	Survival strategy (selling assets, land, or livestock)	Sell agricultural goods	Sell agricultural goods	Credit from bank or local NGO	Sell handicrafts	–	–	–	–

Means different adaptive practices at community level in both irrigated and nonirrigated areas.

of health needs during disasters. More importantly, community health care services are at the front line of protecting health in emergencies because they have knowledge of local risks, which is used to address the actual needs of the community. In a drought-affected region, a health facility at the village as well as the union level is almost absent. Therefore, farmers in both villages pointed out that it is of utmost important to establish a community health care service in their village through which they not only get proper treatment but also receive various health-related information to help them cope with drought.

Vegetable gardening is another suitable adaptive practice for communities. Depending on rainfall variability, the community can select vegetables that need little water. Moreover, they can employ drip irrigation to help avoid water loss. Vegetable gardens not only provide fresh vegetables to meet the nutritional demand of the household but also provide employment to the community during drought periods. Moreover, it increases the ability of the community to generate income from the sale of vegetables and provides other activities during drought time.

These types of adaptation-related studies from different drought-affected regions of the world show similar results. For example, constructing a reservoir, managing irrigation time, constructing canals, changing irrigation systems, and using drought-resistant varieties were deemed by Iranian farmers to be effective adaptive strategies to cope with drought (Marzieh, Ezatollah, & Ali, 2010).

Timeframe and Stakeholder's Role for Implementing Drought Adaptation Practices

To identify the timeframe and stakeholder's role for each DAP, one local-level workshop was arranged with the participation of various stakeholders from different levels. The key aim of this workshop was to harmonize the DAP and to state the role of stakeholders, along with specific

Table 4. Topmost Drought Adaptation Practices at the Community Level.

Rank	Name of the Adaptation Option	Name of the Adaptation Practices
1	Fruit tree plantation	Establishing mango orchard
2	Primary health care facility	Community health care service
3	Drought-tolerant crop	Vegetable gardening

timeframes, which can result in enhancing the drought resilience of farmers in both irrigated and nonirrigated villages of each upazila in the Rajshahi and Chapai Nawabganj districts. In this process, participants were categorized as irrigated and nonirrigated groups to collect their opinions about different DAP. Tables 5 and 6 show that the irrigated and nonirrigated groups mentioned specific timeframes under various actions that came from the farmers' level. The irrigated and nonirrigated groups agreed with 18 short-term actions from the perspectives of both irrigated and nonirrigated areas among 30 adaptive practices at community level. Similarly, they recognized 3 midterm and 9 long-term adaptive practices for both irrigated and nonirrigated areas. These tables also mention the stakeholders' involvement for the successful implementation of DAP. For implementing 6 and 7 DAP in both irrigated and nonirrigated areas, it is also shown in Tables 5 and 6 that the national government was designated as a primary stakeholder whereas, the community was also assigned as primary stakeholder for implementing 8 and 7 DAP in the two areas, respectively. Local government as a primary stakeholder was mentioned for only one adaptive practice – that is, extension worker in both areas.

Effectiveness of Drought Risk Management Policy and Actions

To understand the policy issues regarding droughts, one national-level workshop was organized to explore the possibility of formulating a national-level drought risk management policy in participation with diverse stakeholders.

During the workshop, group discussions among various officials from different departments elicited a number of options. These options aim to make farmers' adaptive practices more successful in the context of Bangladesh. The options are presented in Table 7, which helps irrigated and nonirrigated farmers to find a way to overcome barriers, as mentioned above. Moreover, these options are a list of policy actions that need to be implemented at various levels. These ultimately help to initiate drought risk management policy more effectively and overcome farmers' barriers as well as enhance farmers' adaptive practices and institutional resilience to drought.

From the policy perspective, the institutional provision of support and resources are important for enhancing farmers' adaptive practices. Therefore, considering the options, the policy and actions can be further strengthened through the following suggestions, which will finally help to

Table 5. Timeframe and Stakeholder's Role for Each DAP in Irrigated Areas.

Timeframe				Name of the Stakeholder	
Recommended DAP	Short term	Midterm	Long term	Primary stakeholder	Secondary stakeholder
Received information from TV, radio	[dotted box]			National government	Local government Community
Establish dairy farm		[vertical-hatched box]		Community	Local government NGO, bank, individual
Community health care service	[dotted box]			National government	Local government NGO, community
Dairy farm			[diagonal-hatched box]	Community	NGO, bank
Save money	[dotted box]			Community	NGO, bank
Sell agricultural goods	[dotted box]			Community	NGO
Public awareness program	[dotted box]			National government	NGO
Extension worker	[dotted box]			Local government	National government Community, NGO
Cash as a support of aids from institution			[diagonal-hatched box]	National government	NGO
Use of diesel for irrigation purposes	[dotted box]			Community	National government Local government NGO, bank
Drag the river and use of river water			[diagonal-hatched box]	National government	Local government Community, NGO
Establish drought information center		[vertical-hatched box]		National government	NGO, Community
Use of plastic pipe for irrigation	[dotted box]			Community	Local government NGO, individual
Vegetable gardening	[dotted box]			Community	Research institute NGO, individual
Establish mango orchard			[diagonal-hatched box]	Community	Research institute NGO, bank, individual

[dotted box] Practices that can be implemented during drought period. [vertical-hatched box] Practices that can be implemented in 2–3 years. [diagonal-hatched box] Practices that can be implemented in 3–5 years.

formulate drought risk management policy and actions for northwestern Bangladesh.

Improvement of Water Resource Management Policy. In Bangladesh, the National Water Resources Council (NWRC) is the highest national body for the formulation of water policy. The National Water Policy, formulated in 1999, provides guidelines that cover agriculture, fisheries, industry, navigation, the environment, basin-wide planning, water rights and allocations, public and private investment, water supply, and sanitation. The policy underscores the broad principles of water resource development and its rational utilization. It emphasizes both private and public actions and highlights the importance of the conjunctive use of ground and surface water.

Recently, the management of water resources has become a crucial need in Bangladesh because of growing demand for water and increasing conflict over its alternative uses. As the population increases and makes various uses of water, the growing scarcity of water becomes a serious issue in Bangladesh. In addition, climate change brings severe changes in water-related issues in Bangladesh, which are further exacerbated by precipitation and temperature changes. Too little water in the dry season and too much during the monsoon season make Bangladesh more vulnerable to various water-related issues. Water can be both scare and abundant in Bangladesh. But during the dry season, there is a critical shortage of water, hampering irrigation, fishing, and navigation activities. An increasing demand for both ground and surface water for irrigation purposes occurs in the dry season. Considering all these issues, an improved water resource management policy is required that will consider less water-sensitive crops, dredging of rivers and conserving water in the water bodies during the rainy season, and proper utilization of water.

Resource Allocation. Bangladesh experiences natural calamities almost every year. With the support of various development partners, government of Bangladesh (GoB) has established budgetary allocations and dedicated funds for climate change adaptation (CCA), climate change mitigation (CCM), and DRR. But recognized limitations such as financial resources and/or operational capacities hinder substantial achievements for Bangladesh. As compared with CCA, the total DRR budget is now about 4.5% of the national budget. On the other hand, in aspects of CCA, GoB has declared a climate change fund with a total annual allocation of 100 million USD per year since 2009–2010, and the same has been kept for the year 2010–2011.

Table 6. Timeframe and Stakeholder's Role for Each DAP in Nonirrigated Areas.

Timeframe				Name of the Stakeholder	
Recommended DAP	Short term	Midterm	Long term	Primary Stakeholder	Secondary stakeholder
Received information from TV, radio	■			National government	Individual
Community savings and credit system			■	Community	Individual
Community health care service	■			National government	NGO, community, individual
Dairy farm			■	National government	Individual
Save money	■			Community	Individual
Sell agricultural goods	■			Community	Individual
Extension worker	■			Local government	National government, NGO
Public awareness program	■			National government	
Electricity as a support of aids from institution			■	National government	Local government, bank
Use of diesel for irrigation purposes	■			Community	National government Local government
Drag the river and use of river water			■	National government	Local government
Establish drought information center			■	National government	Research institute Individual
Use of plastic pipe for irrigation	■			Community	National government Local government, NGO
Vegetable gardening	■			Community	National government Research institute, NGO Individual
Establish mango orchard		■		Community	National government Individual

Dotted box: Practices that can be implemented during drought period. Vertically lined box: Practices that can be implemented in 2–3 years. Hatched box: Practices that can be implemented in 3–5 years.

Table 7. Name of the Option for Drought Risk Management Policy and Actions.

Serial No.	Policy Options
1	Improvement of water resource management policy
2	Resource allocation (i.e., budget allocation)
3	Sustainable land management program
4	Emphasis on proper management of rainwater harvesting
5	Planning for cropping zone
6	Re-excavation of ponds/kharis for water restoration
7	Emphasis on the cultivation of short-duration crops

Moreover, Comprehensive Disaster Management Programme (CDMP) of the Ministry of Food and Disaster Management (MoFDM), 3.2 USD million were spent to implement five small-scale community-level risk reduction projects involving NGOs and local-level disaster management committees; these funds have been increased up to 14 million USD for the years 2011–2014. During this time, several activities are underway, such as allocation of 42.5 million USD for the expansion of irrigation and removal of water from waterlogged areas and an agriculture insurance plan of 1.07 billion USD to provide the small and medium farmers with resources during periods of drought.

It is interesting that more development partners are increasing their engagement and allocations for disaster and climate risk reduction. Among them Department for International Development (DFID), Danish International Development Agency (DANIDA), European Commission (EC), Swedish International Development Cooperation Agency (SIDA), government of Norway, Australian government, Japan International Cooperation Agency (JICA), the World Bank and United Nations agencies are giving support to national DRR initiatives. Although the allocated resources increased, they are still far below what is required to reduce the disaster impact to vulnerable communities, as Bangladesh is one of the most highly populated countries experiencing multiple hazards. Therefore, the potential utilization of various resources should be allocated under the need of various ministries and departments to be explored with effective management. Often, lack of coordination in funding allocation (by different government and nongovernment sources) limits the optimization of available resources. Therefore, a coordinated and concerted social safety net program is required.

Sustainable Land Management Practices. Sustainable land management (SLM) is defined as a knowledge-based procedure that helps to integrate land, water, biodiversity, and environmental management (including external inputs and outputs) to meet rising food and fiber demands while sustaining ecosystem services and livelihoods. SLM is necessary to meet the requirements of a growing population. Improper land management can lead to land degradation and cause a significant reduction in production and service functions.

In Bangladesh, land is the base for crop production. It is the most important economic and social asset for many Bangladeshi people, particularly the poor. Soil resources of the country experience tremendous pressure for increased food production. Increased cropping intensity and mineralization of soil organic matter exhausts the soil's capacity to support crops, which enhances severe land degradation. Drought also causes land degradation. As a result, the overall socioeconomic and environmental balance is affected and the livelihood opportunities are also diminished. Drought also causes the loss of productive agricultural lands, deteriorates water and natural resources, diminishes health and nutrition, and increases vulnerability to natural disasters.

To recover the soil as well as land condition, the Soil Resources Development Institution (SRDI) of the Ministry of Agriculture (MoA) is working to improve soil health and preserve it for future generations. The SRDI prepared the Upazila Land and Soil Resources Utilization Guide for 459 upazilas throughout the country, which will help farmers to apply fertilizers according to the fertility status of the soil. Moreover, MoA is working with the Ministry of Land to enact the proposed Agricultural Land Conservation and Land Use Act 2011 to safeguard agricultural lands from encroachments for developments. The MoA is encouraging farmers to use organic fertilizers such as compost and farmyard manure to safeguard soil health.

Various programs relating to soil and land improvement have already been taken by MoA and SRDI; therefore, its effective practices should be assured at the farmers' level. Farmers should care about soil health. On the other hand, the Department of Agricultural Extension, which deals with this soil-testing activity, should continue their practices at the union and upazila levels to assist farmers.

Emphasis on Proper Management of Rainwater Harvesting. Rainwater harvesting refers to the collection and storage of rainwater in situ or within the vicinity of rainfall. At first, rainwater harvesting was practiced for irrigation purposes only. Now it is practiced for drinking and domestic purposes in many countries of the world. In Bangladesh, this practice exists in rural areas mainly for drinking and cooking purposes, serving as an alternative water source, especially in arsenic-, salinity-, and drought-affected areas. But in drought-affected area, the practice is very limited. They only harvest rainwater into their pond and use whenever they need it for their fields. But during the dry season, the limited surface water collected through rainwater does not fulfill the crops' demand. Therefore, groundwater is heavily extracted for irrigation and groundwater depletion continues during summer; it is a major constraint for the development of a dependable water supply system.

Rainwater harvesting will be an alternative for farmers, providing a more viable option than the other technologies. The annual rainfall of Bangladesh is 2,350 mm; however, it is not uniformly distributed in all areas of the country. For instance, the amount of annual rainfall in the northeast is about 4,338 mm, whereas in the northwest, it is around 1,300 mm. Moreover, the distribution of rainfall is unequal through the year. Most of the rainfall occurs from July to October. Therefore, there is a great opportunity to harvest rainwater during the monsoon season and use it during the dry season.

Depending on the scarcity of drinking water, various types of rainwater harvesting models are used at household and community levels. Cement/mortar jars, ferro-cement tanks, and Roller-Compacted Concrete (RCC) ring tanks are used at the household level and constructed underground rainwater harvesting system is used at the community level. In addition to these, there is an indigenous process of rainwater harvesting that is practiced in different part of Bangladesh – construction of do-it-yourself models (motka). While using these models, the hazardous aspects should be kept in mind, such as the contamination of water and mosquito breeding.

Planning for Cropping Zone. The cropping system of a region is decided by a number of soil and climatic parameters, which determine overall agro-ecological settings for nourishment and appropriateness of a crop or set of crops for cultivation. The choice of crops and cropping systems depends on several factors, such as infrastructure, socioeconomic factors, and technological developments. Farmers need to reap profits for their products so that they will be encouraged to grow more crops subsequently.

Re-Excavation of Pond, Khari, and Canal for Water Restoration. In Barind areas, dry spells occur more frequently during the monsoon season as well as the dry season. During the monsoon season, the farmers draw water from a traditional pond, khari, or a natural canal, which has the potential for conserving water and providing water for irrigation and other purposes, during a scarcity of water for supplemental irrigation at critical stages of crop growth. But during the dry season, the surface water is almost empty everywhere, and farmers use groundwater for irrigation purposes. About 13,000 DTW are being used in the Barind area to withdraw underground water daily for irrigating boro crops. The overexploitation of underground water has adversely affected the region's agriculture, irrigation, ecology, biodiversity, weather, environment, and subterranean water levels.

Therefore, it is imperative to conserve water in surface sources such as ponds, kharis, and canals and to use surface water as much as possible, in order to protect against adverse effects. Because there are many derelict ponds in the Barind area, kharis and canals, if these are re-excavated properly, will reserve the water the year-round and limit the pressure on uses of groundwater in drought-prone areas. For this, national or local government planning is necessary to conserve the existing ponds, kharis, and canals to save agriculture, pisciculture, and forestry as well as the environment of the region.

Emphasis on Short-Duration Crop Cultivation. The high Barind Tract is situated in the northwestern part of Bangladesh. The predominant soil is silty loam to silty clay loam with very low organic matter content, high bulk density, and low pH. The rainfall in this area is erratic and low (average, 1,300 mm per annum) and is concentrated from June to September. Aman is the traditional crop grown during the rainy season. After harvesting the aman in Bangladesh, there is often enough stored soil moisture, along with any subsequent rainfall, to support a short-duration crop (Rahman & Biswas, 1995). However, if the rice is harvested early, then plenty of water remains in the soil, which helps to sustain a short-duration crop. Late harvest of aman rice results in surface soil moisture levels that are too low for successful crop production. The most suitable crops for exploitation of residual moisture after the aman rice harvest are mustard, chickpeas, and mung beans. These crops are already being grown in this region to a small extent. Efforts are required to encourage the extensive use of this practice into this region.

CONCLUSION

According to MoEF (2002), various forecast scenarios predict that dry season rainfall may decrease by 37% by 2050, thus increasing the risk of droughts significantly. Though monsoon rainfall is expected to increase by 28%, intermittent dry and wet spells cannot be ruled out. The northwest region is primarily prone to drought, which is likely to become more frequent and intense along with horizontal expansion of climate change. It is also notable here that changes in the timing of drought will result from irregular behavior, unequal distribution of rainfall, and temperature rise. The major impacts would be on agriculture and availability of water resources for agriculture and domestic use, particularly in the dry season.

The economy of Bangladesh is still very much dependent on agriculture; therefore, adaptation to periods of drought is unavoidable. From this study, the farmers living in the irrigated and nonirrigated areas have already undertaken various types of adaptive initiatives using their own local conditions, knowledge, and skills to adapt to drought. But the current level of adaptation is still inadequate to cope with the future challenges of climate change and drought. Most of the adaptive measures in the study area are reactive rather than proactive, autonomous rather than well-planned. Furthermore, this study has shown that farmers in both areas are mainly doing different types of agronomic practices, water harvesting, and water resource exploitation to cope with drought.

Generally, adaptation to climate change requires a combination of various individual responses at the farm level and assumes that farmers have access to alternative practices and technologies available in their area. On the other hand, adaptation can take many forms, and it can be adopted at various levels – farm, industry, or government. This study not only addresses the farmers recommended and prioritized DAP, but it also points out the timeframe and stakeholder's role, which help in implementing these practices. Successful drought adaptation policy requires greater involvement and support from organizations such as governmental, nongovernmental, and private organizations. At the same time, involvement of rural and community voices in drought management planning and implementation are important for the formulation of drought adaptation policy. In this context, this chapter highlights how to formulate and implement DAP of farmers to cope with it in affected areas of Bangladesh.

NOTE

1. During drought period, students have to walk long distances to attend school. This sometimes causes health-related problem and they are unable to attend school thereby enhancing the *dropout* rate.

ACKNOWLEDGMENTS

The financial support provided by the GCOE-ARS program of Kyoto University is highly acknowledged. The first author specially thanks the Graduate School of Global Environmental Studies for supporting her research at Kyoto University, Japan.

REFERENCES

ADB (Asian Development Bank). (2005). *Country strategy and program 2006–2010*. Dhaka, Bangladesh: Bangladesh Resident Mission.

Ahmed, U. A. (2006). *Bangladesh climate change impacts and vulnerability: A synthesis*. Climate Change Cell, Department of Environment Component 4b, Comprehensive Disaster Management Programme, Bangladesh.

Banglapedia. (2006). *Drought in Bangladesh*. Retrieved from http://banglapedia.search.com.bd/HT/D_0284.htm

Bari, M. F., & Anwar, A. H. M. F. (2000). Effects on irrigated agriculture on groundwater quality in Northwestern Bangladesh. In *Proceedings of integrated water resources management for sustainable development*, New Delhi (Vol. I, pp. 19–21).

Climate Change Cell. (2009). Department of Environment. Ministry of Environment and Forests, Bangladesh.

Coleen, V., Laing, M., & Monnik, K. (2006). *A global assessment*. London: Routledge.

Fukuyama, F. (1999). *The great disruption. Human nature and the reconstitution of social order*. London: Profile Books.

Habiba, U. (2012). *Enhancement of drought risk management policy and actions incorporating farmer's adaptive practices in northwestern Bangladesh*. PhD thesis, Kyoto University, Japan.

Habiba, U., & Shaw, R. (2012). Bangladesh experiences of community based disaster risk reduction. In R. Shaw (Ed.), *Community, environment and disaster risk management* (Vol. 10, pp. 91–111). Bingley, UK: Emerald Group Publishing Limited.

Habiba, U., Shaw, R., & Takeuchi, Y. (2011a). Drought risk reduction through a socio-economic, institutional and physical approach in the northwestern region of Bangladesh. *Environmental Hazard, 10*(2), 121–138.

Habiba, U., Shaw, R., & Takeuchi, Y. (2011b). Socio-economic impact of droughts in Bangladesh. In R. Shaw & N. Huy (Eds.), *Droughts in Asian monsoon region: Community, environment and disaster risk management* (Vol. 5, pp. 25–48). Bingley, UK: Emerald Group Publishing Limited.

Habiba, U., Shaw, R., & Takeuchi, Y. (2012). Farmer's perception and adaptation practices to cope with drought: Perspectives from Northwestern Bangladesh. *International Journal of Disaster Risk Reduction*. doi:10.1016/j.ijdrr.2012.05.004

Habiba, U., Takeuchi, Y., & Shaw, R. (2010). Overview of drought risk reduction approaches in Bangladesh. In R. Shaw & J. Pereira (Eds.), *Climate change adaptation and disaster risk reduction: Issues and challenges* (Vol. 8, pp. 37–58). Bingley, UK: Emerald Group Publishing Limited.

Iqbal, A., & Ali, M. H. (2001, April). *Agricultural drought: Severity and estimation*. A report prepared under the project BGD/95/006 under the Ministry of Agriculture, Government of the People's Republic of Bangladesh, Dhaka.

Jones, J. W. (2003). Agricultural responses to climate variability and climate change. Paper presented at Climate Adaptation.net conference "Insights and Tools for Adaptation: Learning from Climate Variability," November 18–20.

Kandlinkar, M., & Risbey, J. (2000). Agricultural impacts of climate change: If adaptation is the answer, what is the question? *Climatic Change*, *45*, 529–539.

Marzieh, K., Ezatollah, K., & Ali, A. (2010). A typology of farmers' drought management. *American-Eurasian Journal of Agriculture & Environmental Science*, *7*(4), 415–426.

MoEF (Ministry of Environment and Forests). (2002). Government of the People's Republic of Bangladesh, Dhaka.

National Drought Mitigation Center (NDMC). (2006). *What is drought? Understanding and defining drought*. Retrieved from http://www.drought.unl.edu/whatis/concept.htm

Nasreen, M., & Hossain, K. M. (2002). Sustainable development: Bangladesh perspective. Paper presented at the NGO session of World Summit on Sustainable Development (WSSD), Johannesburg, South Africa, August 26–September 4.

Paul, B. K. (1998). Coping mechanisms practiced by drought victims (1994/95) in North Bengal, Bangladesh. *Applied Geography*, *18*, 355–373.

Rojas Blanco, A. (2006). Local initiatives and adaptation to climate change. *Disasters*, *30*(1), 141–147.

Shahid, S., & Behrawan, H. (2008). Drought risk assessment in the western part of Bangladesh, natural hazards. *Journal of the International Society for the Prevention and Mitigation of Natural Hazards*, *46*, 391–413.

Shahid, S., & Hazarika, M. K. (2009). Groundwater drought in the northwestern districts of Bangladesh. *Water Resource Management*, *24*(10), 1989–2006.

Tanner, T. M., Hassan, A., Islam, K. M. N., Conway, D., Mechler, R., Ahmed, A. U., & Alam, M. (2007). *ORCHID: Piloting climate risk screening in DFID Bangladesh*. Detailed research report. Institute of Development Studies, University of Sussex, Sussex.

Wilhite, D. A., Hayes, M. J., & Knutson, C. (2000). Planning for drought: Moving from crisis to risk management. *Journal of the American Water Resources Association*, *36*, 697–710.

PART II
ADAPTATION AND RISK REDUCTION STRATEGIES AS LOCAL RESPONSES TO DISASTERS

CHAPTER 7

PEOPLE'S PERCEPTION ON NATURAL DISASTERS AND LOCAL SURVIVAL STRATEGIES IN SUNDARBAN REGION: A STUDY OF GOSABA BLOCK IN SOUTH TWENTY FOUR PARGANAS DISTRICT IN WEST BENGAL, INDIA

Tarun Kumar Mondal

ABSTRACT

Sundarban is the largest active delta in the world lying at the estuaries of the Rivers Ganga and Brahmaputra. This region is rich in biodiversity and declared World Heritage Site by *UNESCO. This delta region is prone to severe natural disasters as well as man-induced catastrophic events. Gosaba Block which is identified as the study area located in Indian part of the Sundarban delta, in South Twenty Four Parganas*

Risks and Conflicts: Local Responses to Natural Disasters
Community, Environment and Disaster Risk Management, Volume 14, 165–184

ISSN: 2040-7262/doi:10.1108/S2040-7262(2013)0000014013

district of West Bengal puts forward an ideal portrait for people's struggle and survival strategies against natural disasters. An attempt has been made in this chapter to study the perception of the main occupational groups on the effects of major natural disasters viz. floods, tidal surges and cyclones in Gosaba Block. An endeavor has also been made to explore local survival strategies which are effective for their sustenance in this vulnerable region. Three-hundred persons from different occupational groups, that is, people engaged in agriculture, fishing, crab collection, tiger prawn seed collection, wood collection and honey collection have been surveyed through a pre-designed questionnaire. Six Focus Group Discussions have been conducted, in which each of the groups comprised six members from a particular occupation. The study has revealed that all the occupational groups perceive threats differently from natural disasters and their perceptions vary according to their levels of exposure to the environment. To cope with the natural disasters, each occupational group has devised distinct survival strategies. For proper management of natural disasters in Sundarban delta region, the perception of people with different livelihoods and their survival strategies should be incorporated.

Keywords: Perception; livelihood; disaster; Gosaba; Sundarban

INTRODUCTION

In a region, an individual or community perceives threats with different intensity from natural disasters and their perception varies according to their levels of exposure to the environment. 'Human beings' perception of environment varying between individuals, groups or communities basically depend on the question of their ability to receive, store and analyse information about the environment and its vagaries' (Mondal, 2013). Wachinger and Renn (2010) have indentified four key factors responsible for the variations in perceptions: the type of risk, the risk context, the personality of the individual and the social context. According to Battista and Baas (2004), 'communities in the high-risk and disaster-prone areas' are familiar with a lot of 'episodes of disaster' and gained substantial 'experience and knowledge of the causes and nature of the recurrent disasters in their areas.' By default, local people have to act immediately when a disaster strikes (Dekens, 2007). Based on their indigenous knowledge and experience, local people take decisions to combat natural disasters in their own way. According to Patterson, Weil, and Patel (2010), 'Most decisions are the products of complex interactions between individuals, their communities, their surroundings and the natural environment itself.' Consequently, these decisions result in distinct survival strategies of individuals or communities depending on their capacity to cope with natural disasters. These local survival strategies are indispensible for the management of natural disasters. Effective management requires a bottom-up approach instead of a top-down approach. Bottom-up approaches follow the line of people's perception of natural disasters in the management which is both participatory and sustainable (Jessamy & Turner, 1999) and emphasizes on local adaptability. In this context, the present study examines the perceptions of the different occupational groups on effects of predominant natural disasters occurring in Gosaba Block of the Indian Sundarban delta and local survival strategies.

SUNDARBAN: A UNIQUE AND MOST VULNERABLE ECOSYSTEM

Sundarban lying at the estuaries of the Rivers Ganga and Brahmaputra is the largest active delta in the world. This delta region covers an area of 25,500 sq.kms. extending across India and Bangladesh. The Indian part of the delta covers an area of 9,630 sq.kms., out of which, 4,264 sq.kms. area is forest land and 5,366 sq.kms area is human habitation. Indian Sundarban comprises of 102 islands, out of which 54 islands are inhabited with a total population of 3,755,924 persons (Census of India, 2001b).

Sundarban portrays truly unique as well as a fragile ecosystem. This largest single mangrove forest block in the world (including Bangladesh) is the only habitat for Royal Bengal Tiger and has been recognised as the largest 'carbon sink' in South Asia. Sundarban is rich in biodiversity and most of the terrestrial and aquatic species are endemic and endangered here. Naturally, this biodiversity rich region has been declared as *World Heritage Site* by UNESCO.

This delta region is prone to severe natural disasters. This vast mangrove forests protect land, people and life from the damaging effects of severe cyclones and tidal surges. It has been estimated that 8 million people (including India and Bangladesh) directly depend on Sundarban for their sustenance and most of them are below poverty line (Chowdhury, Mondal, Brahma, & Biswas, 2008). Singh, Bhattacharya, Vyas, and Roy (2010) opines that the ecological significance of Sundarbans is a well-established fact. However, the issues of economic significance and dependency are less addressed in the existing literature.

The fragility of this ecosystem of Sundarban is being aggravated by the dangers of climate change with increasing probability of more disasters (WWF-India, 2010). 'Frequent climatic insult is a regular feature – cyclonic storm; inrush of tidal waves and flooding is the cause of recurrent damage of life, crops and property every year' (Chowdhury et al., 2008). Over a few decades, increase in temperature at an alarming rate has posed a great threat not only to the future existence of human habitation in this region but also to the existence of this entire productive ecosystem. Mitra, Gangopadhyay, Dube, Schmidt, and Banerjee (2009) observe that over the last three decades, the surface water temperature has been rising at the rate of 0.5 degree Celsius, whereas the average global warming rate is 0.06 degree Celsius, and that Sundarban is becoming a most horrible spot for climate change. It has also been estimated that 28% of the mangrove ecosystem was lost due to sea-level rise over the last 40 years (Kaeslin, Redmond, & Dudley, 2012).

Hence, the study on the perceptions of local people about the adverse effects of natural disasters and their adaptive strategies for survival in this vulnerable region is highly topical from the point of the sustainable management of natural disasters.

Study Area: Gosaba Block – The Hamilton's Island

Gosaba Block, the study area, is located in Indian part of the Sundarban delta, in South Twenty Four Parganas district of West Bengal. This block is one of the most vulnerable and disaster-prone blocks in Sundarban and puts forward an ideal portrait for people's struggle and survival strategies against natural disasters. Human habitation, in this southernmost Block of Sundarban, is adjacent to the deep forests. Inhabitants of this block are compelled to use land, forest and water resources for their sustenance (Fig. 1).

Gosaba Block is also known as Hamilton's Island. The history of human habitation in the colonial period in Gosaba Block evolved a distinct culture. Sir Daniel Hamilton, a Scottish businessman, took lease of the three uninhabited islands: Gosaba, Rangabelia and Satjelia (presently under Gosaba Block) from the British Government in 1903. He 'wished to implement his idea of cooperative community living in a self-sustained way in the deltaic islands of Sundarbans' (District Human Development Report,

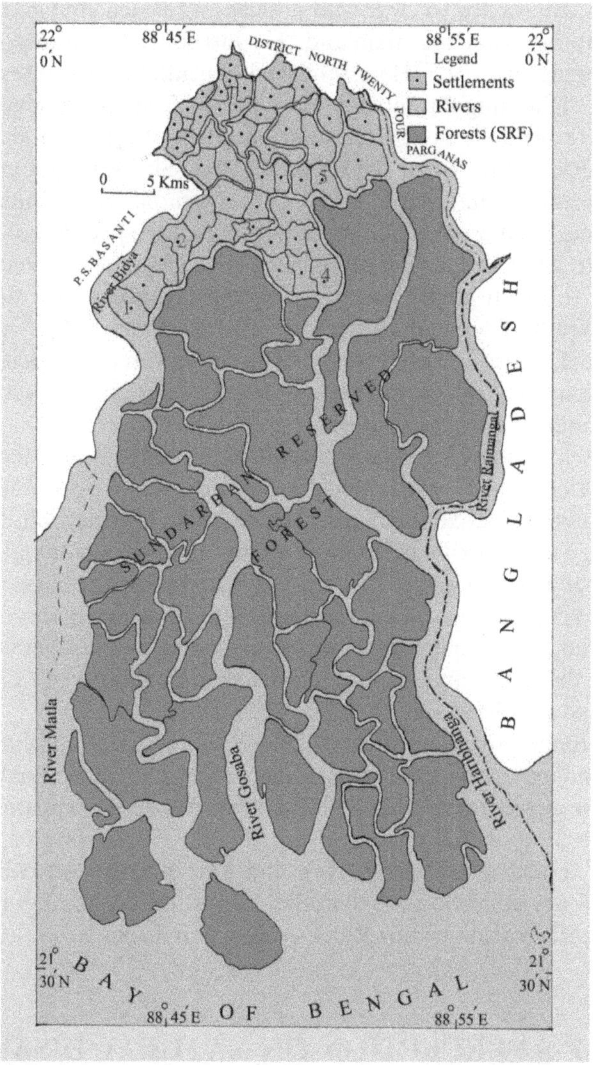

Fig. 1. Location of Gosaba Block. Location of sample villages: 1. Mathurakhanda, 2. Bijoynagar, 3. Pakhiralay, 4. Lahiripur, 5. Kalidaspur. *Source*: Modified from West Bengal Administrative Atlas, Census of India (2001a).

South 24 Parganas, 2009). Sir Hamilton established '*Bengal Young Men's Zamindary Co-operative Society*' and he was recognised as the pioneer in the history of co-operative movement in India. The present day human habitation in this area is an outcome of his monumental success of land reclamation. He started to clear wild jungles and raised embankments using imported labour from the mainland. He introduced rainwater harvesting through digging of ponds. He established educational institutes and medical centres, Dharamgola (grain reservoir) and also introduced *Gosaba Currency*. He brought in salt-resistant varieties of paddy and established the Jamini Rice Mill. Misra (2007) stated that, Sir Hamilton created such an environment in this island wherein 'harmonious inter-ethnic relations' was established and a sense of oneness among all the people was developed. According to him, 'this feeling of oneness as an island shelter has also influenced them to live together and fight together against all the natural odds like flood, wrath of sea and all the fear of wild animals.' As a result, in the Gosaba Block, this distinct history of social awareness, social integrity, communal harmony to combat natural disasters has given a new dimension to the inhabitants' survival strategies.

This present study of people's perception on natural disasters and local survival strategies in Gosaba Block is divided into two sections. The first section involves the analysis of people's perception of natural disasters. In this section, variations in perception of the main occupational groups on the effects of major natural disasters in five selected villages in Gosaba Block have been analysed, assuming that nature of occupation determines the social and environmental awareness of a person. The following section investigates the local survival strategies of the inhabitants of Gosaba Block to combat natural disasters.

Gosaba Block covers an area of 296.73 sq.kms. and consists of 50 villages. According to the Census of India (2001b), total population of this block has been recorded as 222,822 persons. Work participation rate has been reported as 38.62%, out of which 73.53% workforce is directly engaged in agriculture. Fishing, crab and tiger prawn seed collection and wood and honey collection are the other important occupation in the study area. Basic information of this Block is shown in Table 1.

PEOPLE'S PERCEPTION ON NATURAL DISASTERS: A STUDY FROM FIVE SELECTED VILLAGES IN GOSABA BLOCK

This study on people's perception about the adverse effects of natural disasters across the main occupational groups in the selected five villages is based on most severe natural disasters: (i) floods (F), (ii) tidal surges (TS)

Table 1. Gosaba Block: Some Basic Information at a Glance.

Total area	296.73 sq.kms.
Total population	222822 Persons
Male population	113913 Persons
Female population	108909 Persons
SC population	64.28%
ST population	9.23%
Literacy rate	68.93%
Male literacy rate	80.74%
Female literacy rate	56.60%
Work participation rate	38.62%
Agricultural worker	73.53%
Total families	46962
Families below poverty line (BPL)	19853 (42.27%)
Families above poverty line (APL)	27109 (57.73%)

Sources: (i) Census of India, 2001b, West Bengal, Series 20, Volume-1, Primary Census Abstract. (ii) Block Development Office, Gosaba, South 24 Parganas District.

and (iii) cyclones (C). Six main occupational groups, selected for the perception study are (i) cultivators (CU), (ii) agricultural labourers (AL), (iii) large fishermen (LF), (iv) small fishermen (SF), (v) crab and tiger prawn seed collectors (CTS) and (vi) wood and honey collectors (WHC).

Five villages – Mathurakhanda, Bijoynagar, Pakhiralay, Lahiripur and Kalidaspur – have been selected for perception study about the adverse effects of natural disasters. Selection of these sample villages were based on three distinct criteria: (i) these five villages are severely prone to floods, tidal surges and cyclones; (ii) all the selected six occupational groups exist in these villages and (iii) location of all these villages are adjacent to the Sundarban Reserve Forest and well-exposed to water bodies. Due to these specific location factors, people of these villages earn their livelihood from forest and water resources along with land resources. Hence, people's interface with the environment is complex and multidimensional.

For this study, field survey has been conducted during March, 2011. Ten persons from each occupational group from each village have been surveyed through pre-designed questionnaires and altogether total number of respondents was 300 persons. Six Focus Group Discussions have been conducted, in which each of the group comprised of six members from a particular occupation.

The intensity of the perceived threat from any particular natural disaster has been measured through a 7-point scale (Korpela & Hartig, 1996).

- Initially, Individual score (IS), ranging from 0 to 6, have been calculated based on the given score by the each respondent for the perceived threat from each disaster. For each disaster per occupation group, Village-wise Score (VS) has been calculated on the basis of the individual score. Village-wise Score (VS) has been summed up to find out the Total Score (TS) for each natural disaster per occupation group.

$$VS_{FC} = \sum_{i=1}^{10} IS_{FCi} \quad (1)$$

VS_{FC} = Village-wise score of a particular hazard, F = flood for a particular occupation group, and C = cultivator. IS_{FC} = Individual score of a particular hazard, F = flood for a particular occupation group, and C = cultivator.

$$TS_{FC} = \sum_{i=1}^{5} VS_{iFC} \quad (2)$$

TS_{FC} = Total Score of a particular hazard e.g. F = flood, for a particular occupation group e.g. C = cultivator.

- Finally, Total Score (TS) for each disaster has been transformed into an index value. Index value ranges from 0 to 1 (0 for minimum possible total score and 1 for maximum possible total score) (Table 2 and 3).

VILLAGE WISE VARIATIONS OF PERCEPTION ON NATURAL DISASTERS

Cyclone has been perceived as the most threatening natural disaster compared to flood and tidal surge across all villages. This area was severely devastated by the cyclone *AILA* on 25th May 2009. This awful memory is still alive in the minds of the inhabitants of this region. Analysing the Village-wise scores for each natural disaster, Mathurakhand Village has been found as the most vulnerable to all three natural disasters and Kalidaspur Village the least vulnerable. Socio-economic differentiations of

Table 2. Occupation Wise Total Scores (TS) of the Perception about the Adverse Effects of Natural Disasters Using a 7-Point Scale in Gosaba Block.

Occupational Categories	Natural Disasters		
	Floods	Tidal surges	Cyclones
Cultivators (CU)	250	209	213
Agricultural labourers (AL)	221	199	229
Small fishermen (SF)	159	205	241
Large fishermen (LF)	92	101	275
Crab and tiger prawn seed collectors (CTS)	210	208	235
Wood and honey collectors (WHC)	207	186	234
Total	1139	1108	1427

Source: Computed by the author.

Table 3. Occupation Wise Index Values of the Perception about the Adverse Effects of Natural Disasters using a 7-Point Scale in Gosaba Block.

Occupational Categories	Natural Disasters		
	Floods	Tidal surges	Cyclones
Cultivators (CU)	0.83	0.70	0.71
Agricultural labourers (AL)	0.74	0.66	0.76
Small fishermen (SF)	0.53	0.68	0.80
Large fishermen (LF)	0.31	0.34	0.92
Crab and tiger prawn seed collectors (CTS)	0.70	0.69	0.78
Wood and honey collectors (WHC)	0.69	0.62	0.78

Source: Computed by the author.

the inhabitants as well as geographical location can be attributed for these regional variations. Mathurakhand Village is located at the southernmost end of the habitat areas of Gosaba Block, surrounded by wide rivers on three sides and due to its geographic location, cyclones and tidal surges directly hit the inhabitants of this village. On the contrary, the location of Kalidaspur Village is on the northernmost extreme among the selected villages and surrounded by less wide rivers compared to Mathurakhand Village. Apart from the differences in geographical locations, socio-economic condition of the Mathurapur Village is backward compared to Kalidaspur Village. (Table 4–8 and Figs. 2–6)

Table 4. Village-Wise Score (VS) of the Perception about the Adverse Effects of Natural Disasters Using a 7-Point Scale in Pakhiralay Village.

Occupational Categories	Natural Disasters		
	Floods	Tidal Surges	Cyclones
Cultivators (CU)	51	40	40
Agricultural labourers (AL)	47	39	45
Small fishermen (SF)	35	42	49
Large fishermen (LF)	22	21	56
Crab and tiger prawn seed collectors (CTS)	44	42	48
Wood and honey collectors (WHC)	45	36	48
Total	244	220	286

Source: Field survey (March 2011).

Table 5. Village-Wise Score (VS) of the Perception about the Adverse Effects of Natural Disasters Using a 7-Point Scale in Mathurakhanda Village.

Occupational Categories	Natural Disasters		
	Floods	Tidal surges	Cyclones
Cultivators (CU)	56	45	45
Agricultural labourers (AL)	50	42	50
Small fishermen (SF)	32	42	53
Large fishermen (LF)	21	18	58
Crab and tiger prawn seed collectors (CTS)	48	43	50
Wood and honey collectors (WHC)	43	40	47
Total	250	230	303

Source: Field Survey (March 2011).

Table 6. Village-Wise Score (VS) of the Perception about the Adverse Effects of Natural Disasters Using a 7-Point Scale in Lahiripur Village.

Occupational Categories	Natural Disasters		
	Floods	Tidal surges	Cyclones
Cultivators (CU)	52	42	47
Agricultural labourers (AL)	45	37	51
Small fishermen (SF)	32	43	45
Large fishermen (LF)	18	24	57
Crab and tiger prawn seed collectors (CTS)	41	44	51
Wood and honey collectors (WHC)	44	39	51
Total	232	229	302

Source: Field Survey (March 2011).

Table 7. Village-Wise Score (VS) of the Perception about the Adverse Effects of Natural Disasters using a 7-Point Scale in Bijoynagar Village.

Occupational Categories	Natural Disasters		
	Floods	Tidal surges	Cyclones
Cultivators (CU)	48	43	42
Agricultural labourers (AL)	41	45	42
Small fishermen (SF)	30	38	50
Large fishermen (LF)	16	20	54
Crab and tiger prawn seed collectors (CTS)	42	42	44
Wood and honey collectors (WHC)	38	36	46
Total	215	224	278

Source: Field Survey (March 2011).

Table 8. Village-Wise Score (VS) of the Perception about the Adverse Effects of Natural Disasters Using a 7-Point Scale in Kalidaspur Village.

Occupational Categories	Natural Disasters		
	Floods	Tidal surges	Cyclones
Cultivators (CU)	43	39	39
Agricultural labourers (AL)	38	36	41
Small fishermen (SF)	30	40	44
Large fishermen (LF)	15	18	50
Crab and tiger prawn seed collectors (CTS)	35	37	42
Wood and honey collectors (WHC)	37	35	42
Total	198	205	258

Source: Field Survey (March 2011).

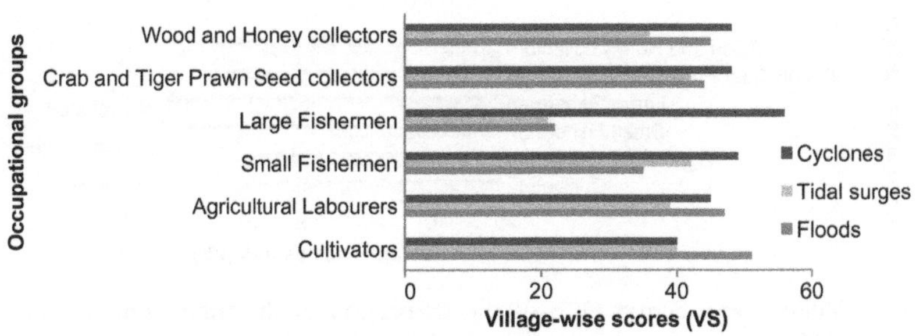

Fig. 2. Village-Wise Scores (VS) of the Perception of the Inhabitants about the Adverse Effects of Natural Disasters in Pakhiralay Village (Based on 7-Point Scale).

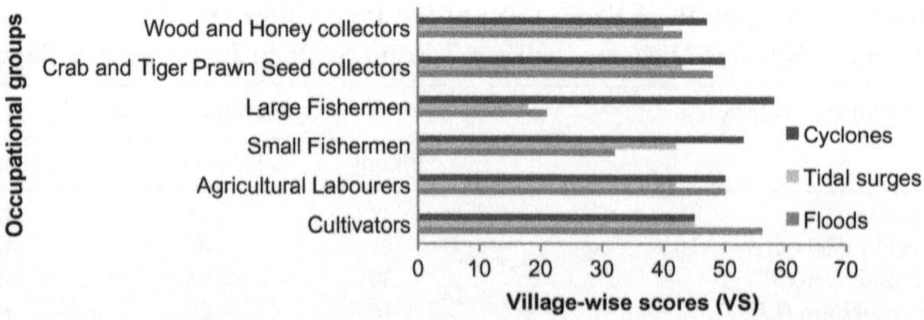

Fig. 3. Village-Wise Scores (VS) of the Perception of the Inhabitants about the Adverse Effects of Natural Disasters in Mathurakhanda Village (Based on 7-Point Scale).

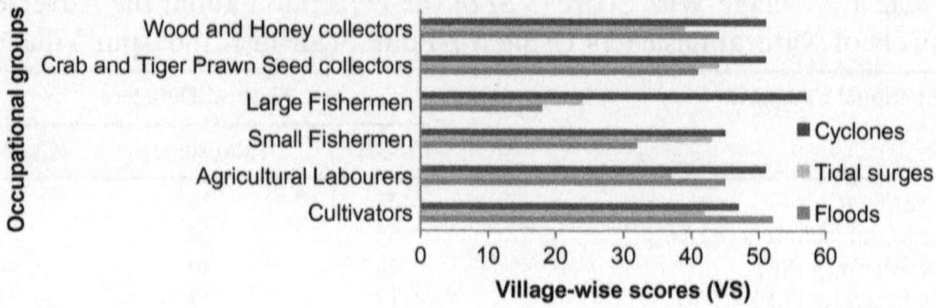

Fig. 4. Village-Wise Scores (VS) of the Perception of the Inhabitants about the Adverse Effects of Natural disasters in Lahiripur Village (Based on 7-Point Scale).

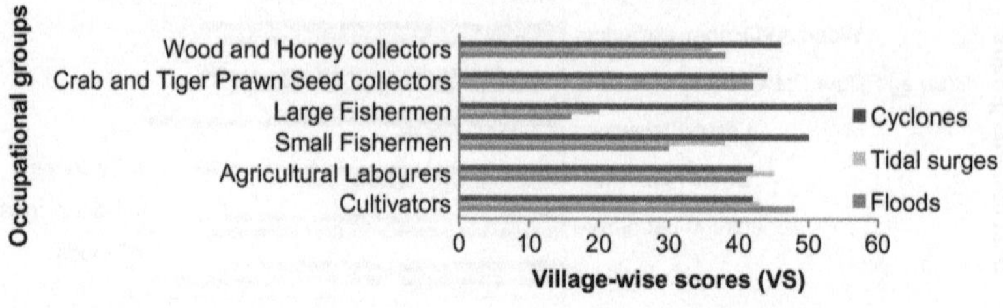

Fig. 5. Village-Wise Scores (VS) of the Perception of the Inhabitants about the Adverse Effects of Natural Disasters in Bijoynagar Village (Based on 7-Point Scale).

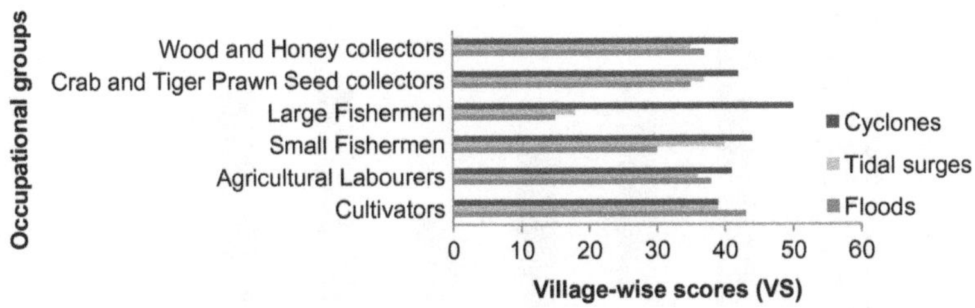

Fig. 6. Village-Wise Scores (VS) of the Perception of the Inhabitants about the Adverse Effects of Natural Disasters in Kalidaspur Village (Based on 7-Point Scale).

OCCUPATION-WISE VARIATIONS ON PERCEPTION OF NATURAL DISASTERS

Floods (F)

Floods are the most common natural disasters in Gosaba Block. Flooding in Gosaba is either caused by excessive rain or by tidal surges and cyclones. Every year, heavy monsoon rain inundates low-lying areas; this flood is less disastrous. The floods caused by storm induced tidal surges and cyclones wreak more havoc.

The index value of the perception about the adverse effect of floods for cultivators is maximum, that is, 0.83, followed by 0.74 for agricultural labourers. The immediate impact of flood is on the standing crops. Flood not only damages the standing crops; submergence of the agricultural land with flood water enhances soil salinity. As a result, agricultural lands become unproductive which directly affects the livelihoods of the cultivators and agricultural labourers. The index values for crab and tiger prawn seed collectors and honey and wood collectors are 0.70 and 0.69 respectively. The fishermen are the least affected group by the floods and the index values for small and large fishermen are 0.53 and 0.31 respectively (Table 3 and Fig. 7).

Tidal Surges (TS)

Gosaba Block is highly vulnerable to tidal surges, which have wreaks devastations especially to the economically marginalised people in past. Tidal surges are always associated with strong winds, storms and cyclones.

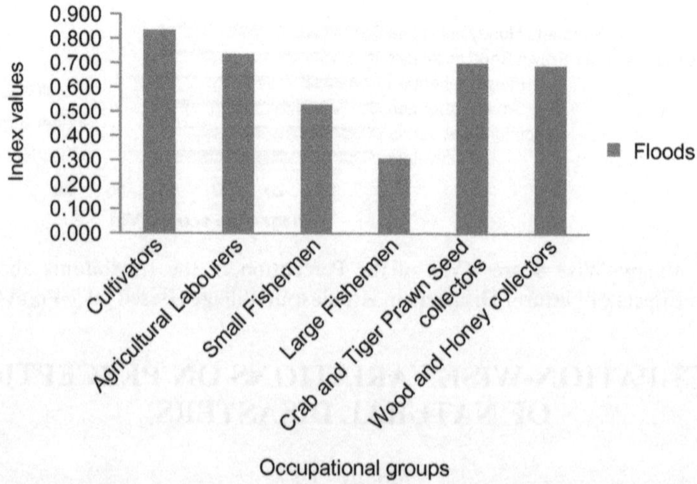

Fig. 7. Index Values of Different Occupational Groups for Floods.

The high tidal surges demolish the embankments and cause tidal floods, resulting in agricultural fields being inundated with saline water. Tidal surges essentially affect the economic activities occurring near the coastal areas. High amplitude tidal surges occur during April–October.

Tidal surges primarily hit the economic activities near the immediate vicinity of the coast. It destroys crops, increases salinity in agricultural lands and imposes maximum threat towards crab and tiger prawn seed collectors and small fishermen. The index values for cultivators, tiger prawn seed collectors and small fishermen are 0.70, 0.69 and 0.68 respectively. The index value for large fishermen is 0.34, that is, lower than the small fishermen. The large fishermen catch fish in the deep sea zone where the effects of the tidal surges are negligible. They are also well-equipped to combat the tidal surges. The index values for agricultural labourers and wood and honey collectors are 0.66 and 0.62 respectively (Table 3 and Fig. 8).

Cyclones (C)

As already stated, Gosaba Block is prone to severe cyclones. Apart from the huge loss of lives and property due to severe cyclone, it generates storm surges and raises the height of sea water level up to 2–3 metres. This storm

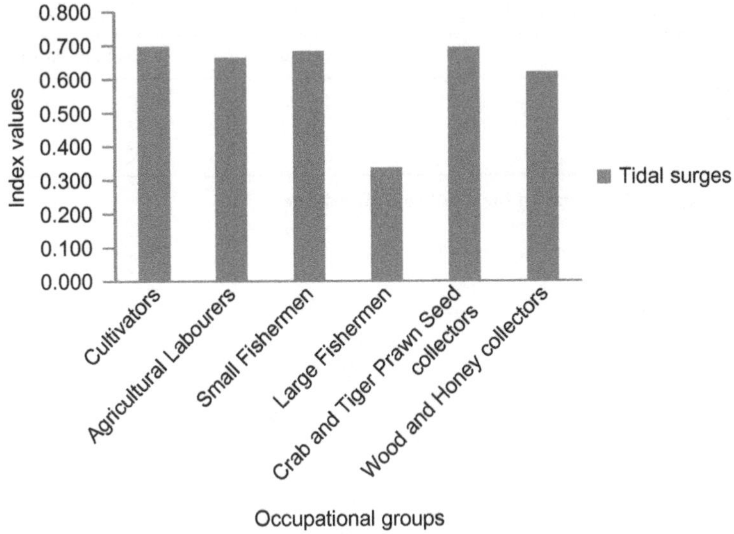

Fig. 8. Index Values of Different Occupational Groups for Tidal Surges.

surges reduce crop lands and spoil the sources of drinking and irrigated water with intrusion of saline water from sea. Fishermen are largely hit by storm surges. Generally, May–November are the most cyclone-prone months. However, even on other months, occurrence of severe cyclones is now quite a common phenomenon in this region. It has been found from a study, over the period from 1970 to 2000, the frequency of storms and cyclones over the Bay of Bengal has decreased but the magnitude of the cyclones has increased remarkably (WWF, 2010). Over the Bay of Bengal, during the last 120 years, recorded severe cyclones have increased by 26% (Singh, 2007). Cyclonic data of the *Cyclone eAtlas,* Indian Meteorological Department, reveal the increasing trend of severe cyclonic storms from 1981 to 2010 (Table 9). Cyclones Sidr (2007), Bijli (2008), Nargis (2008) and Aila (2009) are the most devastating, severe cyclones in the recent past.

Cyclone hits across all the occupational groups. However, the perceived threat from cyclones is highest for large and small fishermen. The index values of the small and large fishermen are 0.80 and 0.92 respectively. The small fishermen, who catch fish near the coast, have better chances to come back before the cyclone hit, than large fishermen. For cultivators, agricultural labourers, tiger prawn seed collectors and wood and honey collectors, the index values range from 0.71 to 0.78 (Table 3 and Fig. 9).

Table 9. Average Frequency of Severe Cyclones in Sundarban.

Period	Frequency of Severe Cyclonic Storms/Year
1891–1930	1.6
1931–1970	1.98
1971–2010	2.05

Source: Cyclone eAtlas, Indian Meteorological Department.

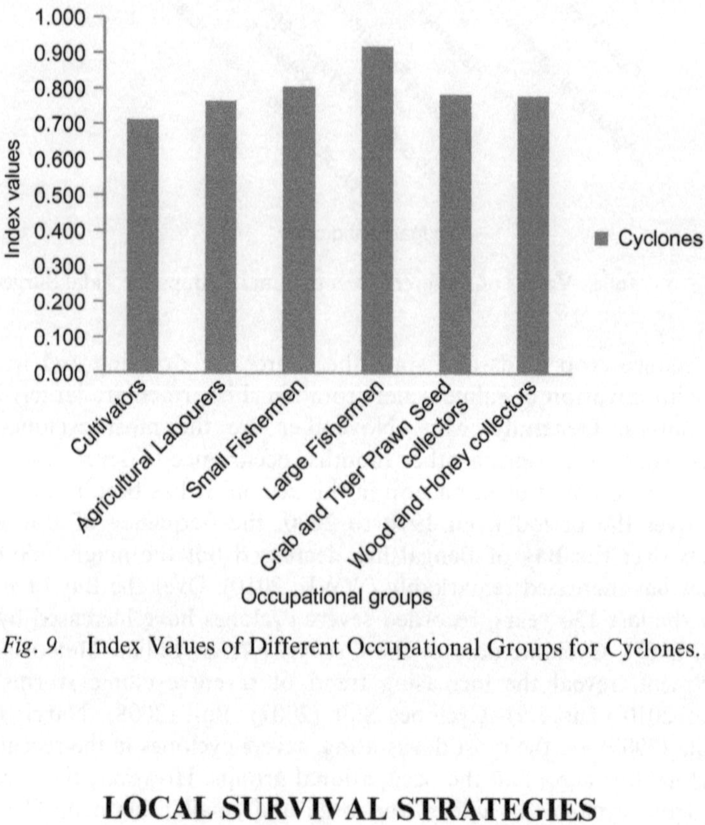

Fig. 9. Index Values of Different Occupational Groups for Cyclones.

LOCAL SURVIVAL STRATEGIES

On the basis of the responses from different occupational groups and focus group discussions, the following local survival strategies have been outlined.

Migration of Adult Work Force

Agricultural activities become less profitable due to increase in salinity in the agricultural fields and it reduces the employment opportunity from agriculture. To maintain their survival, adult workforce often migrates as unskilled labourers. This trend of migration lessens after heavy monsoon rains for few consecutive years since intense rainfall reduces the salinity of the soils.

Plantation of Coconut Trees

Cyclones invite tidal surges and floods. The occurrence of these three disasters results in scarcity of drinking water. Green coconuts are used as alternative source of drinking water and local people emphasize on the plantation of coconut trees.

Plantation of Trees around the Houses

To protect the house from the cyclones and tidal surges, local people planting trees around their houses is a common practice. These trees act as a shield to their houses and minimise the devastating effects of cyclone and tidal surges.

Special Design of Houses

Houses are constructed on the elevated land to prevent inundation during floods and tidal surges. The houses are also designed to allow sufficient ventilation through which winds can pass easily. As a result the gale effects of cyclone on house are reduced.

Use of Traditional Wisdom for Prediction of Cyclones

People can predict cyclone using different traditional wisdom. The common belief is that with approaching cyclones, the number of fish increases, water temperature of river and sea rises, colour of water in the river and sea usually darkens, Cheowa fish begins jumping, breathing problems increases, wind temperature reverses and insects leave their nest and go to elevated places.

Local strategies in Agricultural Practices

Adoption of Salt Tolerant Varieties of Paddy
Paddy is the principal crop produced in this area. Salinity is the main problem for agriculturists. Farmers largely adopt various salt-tolerant paddy varieties, namely, Marichshal, Talmugur, Hoogla, Lilabati, Kumragore, Lalgetu and Kutepatnail.

Dhibi Cultivation
Farmers dig ponds and canals for rainwater harvesting. Soils from the dug ponds and canals are laid on the agricultural fields for raising elevation. These elevated fields are more conducive for crop cultivation in this low-lying region. Agricultural fields are also divided into elevated rows for raising vegetation. The rain water is used for irrigation and these water bodies are also used for pisciculture.

Pot Irrigation
Pot irrigation is a unique technique adopted by the farmers in this region to ensure effective use of fresh water in agriculture. Earthen pots filled with fresh water are kept at the base of the vegetable and fruit-plants. At the bottom of these pots there are holes through which fresh water drip down to the root of the plants.

Embankments and Community Participation

Embankments are considered the 'lifeline' in the history of human habitation in the Sundarban region. Embankments protect human habitat from floods and tidal surges. Active community participation plays a pivotal role in the construction and maintenance of these embankments.

FOLK TRADITION

A unique folk tradition persists among the inhabitants of the study area. In this natural disaster-prone region, this unique folk tradition endows the local community with immense willpower and social strength during natural calamities. The mythic tales of Bonbibi, Dakshin Rai, Kasthadevi and Gangadevi are closely connected with the history of the natural coping and survival strategies of the inhabitants in this fragile ecosystem. Bonbibi (the Goddess of Forests) is a local deity who is believed to protect the forest-goers from the attacks of tigers in Sundarban and fulfill the desires of her believers. Another regional deity, Dakshin Rai (the God of Tiger), is worshipped by peoples from both Hindu and Muslim communities. Again most fishermen of this delta region worship their boats as Kasthadevi (the Goddess of Boat) and Gangadevi (the Goddess of Water). Kasthadevi warns fishermen in stressful situation by making sounds and help them change the direction of the boats and destination from evils. Gunin, Mantra, Tabij, Maduli and Rumal also play a significant role in the day-to-day life of Sundarban's people. In brief, to the conflict-ridden contemporary world, the folk tradition of Sundarban delta provides a trusting example of social integrity and communal amity to combat natural disasters.

CONCLUSION

Socio-economic fabric of the Gosaba Block in the biodiversity-rich mangrove forest of Sundarban delta portrays a unique man – environment relation. To survive against all adversities, local people have developed specific strategies for their survival in this disaster-prone region. These strategies have evolved out of their intense interaction with nature through generations and based on their perception of threat from natural disasters. The study reveals nature of occupation highly influences the perception of threats from natural disasters in this block. The human ecology of Gosaba Block suggests that effective management of natural disasters is possible in this vulnerable region by incorporating people's perception and these local survival strategies in the planning and policy making.

ACKNOWLEDGEMENT

The author would like to acknowledge the Department of Science & Technology (DST), Government of India, for providing financial assistance from DST PURSE Programme, University of Kalyani in carrying out this research.

REFERENCES

Battista, F., & Baas, S. (2004). *The role of local institutions in reducing vulnerability to recurrent natural disasters and in sustainable livelihoods development.* Consolidated Report on Case Studies and Workshop Findings and Recommendations, Rural Institutions and Participation Service, FAO, Rome.

Census of India. (2001a). West Bengal Administrative Atlas, Directorate of Census Operations, Government of India.

Census of India. (2001b). West Bengal, Series 20, Volume-1, Primary Census Abstract, Government of India.

Chowdhury, A. N., Mondal, R., Brahma, A., & Biswas, M. K. (2008). Eco-psychiatry and environmental conservation: Study from Sundarban Delta, India. *Environmental Health Insights*, *2*, 61–76.

Dekens, J. (2007). *Local knowledge for disaster preparedness: A literature review.* Kathmandu, Nepal: International Centre for Integrated Mountain Development (ICIMOD).

District Human Development Report. (2009). South 24 Parganas, Development and Planning Department, Government of West Bengal.

Jessamy, V. R., & Turner, R. K. (1999). *Modelling community response and perception to natural hazards: Lessons learnt from Hurricane Lenny.* CSERGE Working Paper EDM 03-06, School of Environmental Sciences University of East Anglia, UK, 1–24.

Kaeslin, E., Redmond, I., & Dudley, N. (Eds.) (2012). *Wildlife in a changing climate.* Rome: Food and Agriculture Organization of the United Nations.

Korpela, K., & Hartig, T. (1996). Restorative qualities of favorite places. *Journal of Environmental Psychology*, *16*, 221–233.

Misra, S. S. (2007). Delta culture of Sundarban: A case study of Gosaba Block, man in biosphere. In D. Mondal (Ed.), *A case study of Sundarban Biosphere Reserve* (pp. 157–190). New Delhi: Gyan Publishing House.

Mitra, A., Gangopadhyay, V., Dube, A., Schmidt, A. C. K., & Banerjee, K. (2009). Observed changes in water mass properties in the Indian Sundarbans (northwestern Bay of Bengal) during 1980–2007. *Current Science*, *97*(10), 1445–1452.

Mondal, T. K. (2013). Environmental hazards and community responses in Sagar Island, India. *Ocean & Coastal Management*, *71*, 73–78.

Patterson, O., Weil, F., & Patel, K. (2010). The role of community in disaster response: Conceptual models. *Population Research and Policy Review*, *29*, 127–141.

Singh, A., Bhattacharya, P., Vyas, P., & Roy, S. (2010). Contribution of NTFPs in the Livelihood of Mangrove Forest Dwellers of Sundarban. *Journal of Human Ecology*, *29*(3), 191–200.

Singh, O. P. (2007). Long-term trends in the frequency of severe cyclones of Bay of Bengal: Observations and simulations. *Mausam*, *58*(1), 59–66.

Wachinger, G., & Renn, O. (2010). *Risk perception and natural hazards.* CapHaz-Net WP3 Report, DIALOGIK Non-Profit Institute for Communication and Cooperative Research, Stuttgart.

WWF-India. (2010). *Sundarbans: Future imperfect climate adaptation report.* New Delhi, India.

CHAPTER 8

IMPACT OF SOCIAL CAPITAL ON LOCAL COMMUNITIES' RESPONSE TO FLOODS IN SOUTHERN POLAND

Jarosław Działek, Wojciech Biernacki and Anita Bokwa

ABSTRACT

This chapter discusses the relationship between the amount of social capital resources and the local response to floods. It contains selected results of a large study on social attitudes towards natural hazards (floods, storm, and landslides) conducted in Southern Poland. The results focus on the actions taken by the inhabitants of six communities of different sizes threatened by floods. Both individual and collective mitigation behaviors are analyzed. Demographic and social characteristics are used to explain differences in local preparedness. Communities were chosen to represent three historical regions with different levels of bonding and bridging social capital. The results indicate the importance of strong ties (bonding social capital) in local flood preparedness and participatory risk management. Weak ties (bridging social capital) seem to be less useful, although their strengthening is necessary in areas with

Risks and Conflicts: Local Responses to Natural Disasters
Community, Environment and Disaster Risk Management, Volume 14, 185–205

ISSN: 2040-7262/doi:10.1108/S2040-7262(2013)0000014014

high migratory flows. We conclude that local risk assessment calls for better knowledge and understanding of local network capacities and their possible use in local flood risk governance strategies.

Keywords: Flood; social capital; social capacities; risk governance; Poland

INTRODUCTION

In the last three decades, the notion of social capital has been introduced and widely discussed in social and economic sciences (Bourdieu, 1986; Coleman, 1994; Granovetter, 1974; Portes, 1998, 2000; Putnam, 2000; Putnam, Leonardi, & Nanetti, 1993; Woolcock, 1998). It is usually defined as networks of social relations among individuals and groups, which are accompanied by norms of trust and reciprocity (van Deth, 2003; Westlund & Adam, 2010). Large numbers of both theoretical and empirical works have proposed social capital as an additional factor which may contribute to the explanation of differences in economic growth, advancement of democratic processes, strengths and weaknesses of local communities, individual entrepreneurship, quality of life, health, etc. (Gittel & Vidal, 1998; Granovetter, 1974; Hyyppä, 2010; Mihaylova, 2004; Paxton, 2002; Putnam, 2000; Putnam et al., 1993; Westlund & Adam, 2010).

Social networks could be also considered as important when investigating social aspects of natural hazards, especially social vulnerability (Cutter, Boruff, & Shirley, 2003). Their development and use before, during, and after natural disasters may translate into local community social capacities. In the field of natural hazards, social capacity is defined as "the context-related ability to decide and to behave successfully in a certain situation in order to anticipate, respond to, cope with, recover from or adapt to the negative impacts of an external stressor (e.g., a hazardous event) as well as to employ the necessary resources" (Kuhlicke et al., 2011, p. 807). Knowledge, motivational, network, economic, and governance capacities are distinguished in the literature (Kuhlicke, Steinführer, Begg, & Luther, 2012).

This chapter presents the impact of network capacities, treated synonymously to social capital, on the response of local communities to floods in Southern Poland. The studies presented below have been conducted in this part of the country, because the differences observed here in resources of social capital within local communities are deeply rooted in historical processes. The recognition of factors controlling variability of social capital and its relations with other types of social capacities is essential for the governance of natural hazards, including effective risk communication and education strategies.

Southern Poland was chosen as the research area also due to a higher flood risk compared to other regions in Poland (Fig. 1). Mountain and foothill areas are the most endangered ones, due to environmental conditions, that is, the presence of steep slopes often built from low-permeability rocks and the effect of the mountains as the orographic barrier.

Southern Poland lies in the Upper Vistula river and the Upper Odra river drainage basins, located in the Carpathian Mts. and in the Sudety Mts. The Carpathian rivers, in comparison with the rivers in other regions

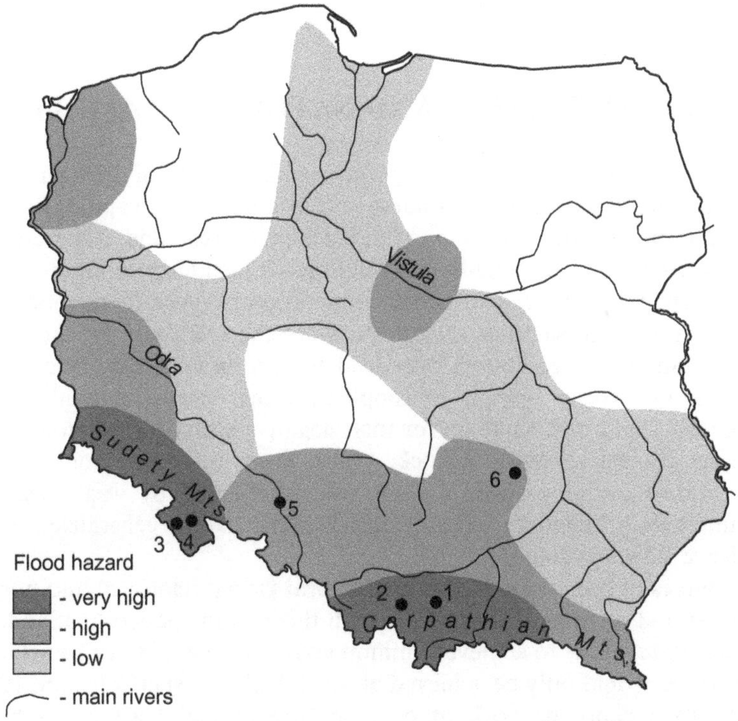

Fig. 1. Flood Hazard Risks in Poland. Explanation: Numbers Correspond to the Localities Included in the Study, see Table 1. *Source*: Data on flood hazard from Maciejewski, 2000.

in Poland, are characterized by the largest irregularity of extreme discharges. In spring, snow thawing contributes to high water levels in the mountain rivers. If the process is extended in time, flood waves are often of a limited size, but if the air temperature rises suddenly, the flood can cause large losses. In summer, intensive thunderstorm precipitation usually results in flash floods with high destructive force but limited spatial range. However, prolonged summer precipitation, linked to the impact of the mountains as the orographic barrier, may cause floods severely affecting large areas (Starkel, 1999). In Poland, 40% of all losses caused by natural hazards are attributed to floods. The frequency and spatial range of flood destruction can vary greatly. For example, during the huge flood of 1997, 52 persons lost their life, about 162,000 of people had to be evacuated and around 45,000 buildings were destroyed (Instytut Meteorologii i Gospodarki Wodnej et al., 2011).

SOCIAL CAPITAL AND NATURAL HAZARDS

Gittel and Vidal (1998), following the seminal work of Putnam et al. (1993), distinguished between bonding and bridging social capital. Bonding social capital includes strong ties and personal trust and it pertains to family, close friends, or neighbourly relations. It facilitates community integration, which should lead to better cooperation between its members and greater readiness to sacrifice for others. Since the sole focus on the closest circle of family or neighbours may lead to amoral familism,[1] the role of "autonomous" social links is to supplement the benefits stemming from "embedded" links and make up for their negative social and economic consequences (Portes, 1998; Woolcock, 1998). These more autonomous relations, that is bridging social capital, are represented by weak ties with acquaintances or members of associations and social (or generalized) trusts (Sander & Lowney, 2006).

Resources of social capital, both structural (social relations) and normative (trust and reciprocity), which flourish through the networks (Sander & Lowney, 2006) allow to achieve common goals which could not be achieved otherwise or would only be achieved at much higher costs (Coleman, 1994). Thanks to a dense network of contacts and mutual trust an exchange between participants of the network of other resources and undertaking joint actions are possible (Gittel & Vidal, 1998). Putnam (2000) underlined the role of associations, in which not only their explicit objectives are important, but also the opportunities created by them allowing people from different groups to interact with each other.

Concerning natural hazards, the key role of network capacity is linked to the ability to exchange resources (especially knowledge) in the case of strong family, friendship, and neighbourhood ties, and also to the trust among the community members (Murphy, 2007). Bonding social capital enables strengthening the memories about earlier events of natural disasters and exchange of information about the possibility of a natural disaster, its mechanisms, and possible mitigation behaviour. This type of social capital may have its negative externalities, when social groups more resourceful in terms of power relations and financial capacities (e.g., local elites, privileged groups) exclude other groups such as minorities, the poorer, older, and less educated from altruistic communities, especially during the response and recovery phase of risk management (Kaniasty & Norris, 1995; Pelling, 1998).

Thus, creating and strengthening of bridging social capital between socially excluded, marginalized groups and the rest of the society are crucial, when dealing with social vulnerability to natural hazards (Adger, Hughes, Folke, Carpenter, & Rockström, 2005; Cutter et al., 2003). Many community-based organizations play a very important role by establishing ties among various social groups. This allows the transfer of knowledge about local natural hazards to, for example, new inhabitants or ethnic minorities which is of high importance in times of intensive mobility (Berke, Kartez, & Wenger, 1993; Bolin & Stanford, 1998). Additionally, bridging social capital in the form of translocal ties may be used by evacuated residents (Elliot, Haney, & Sams-Abiodum, 2010).

Abundant resources of social capital should facilitate taking common decisions (Adger, 2003). Especially the participation of local communities in emergency management helps to develop an active attitude toward formal institutions (Dynes, 2002; Murphy, 2007). Those actions contribute to all phases of the natural hazard risk management cycle, from prevention and preparation to response and recovery. The assessment of social capital resources and their strengthening is particularly important in the planning phase, so as to use them later during the disaster occurrence. However, the existence of social capital does not automatically result in resilience and therefore it is necessary to improve our understanding of the links between various social ties and trust versus actions undertaken by the local community in the context of natural hazards (Murphy, 2007).

These considerations show that strong network capacities (high resources of different types of social capitals) play an important role in social capacity building in general (Kuhlicke et al., 2011) as they contribute to strengthening of other capacities, especially knowledge capacities – through information exchange about hazards and motivational capacities – thanks to the social support and responsibility building within the community. Balanced networks of both bonding and bridging social capital may constitute a determinant of community adaptive capacity (Yohe & Tol, 2002).

The flood risk management policy is implemented in Poland, like in most European countries, according to the EU documents (European Commission, 2003, 2004). The rational management strategies should be based on the following assumptions:

- various actions should be undertaken at all stages of the management process, from prevention, through education to mitigation and recovery after the disaster;
- all potentially endangered persons and legal entities should participate in the management strategy actions;
- the strategy should be realized using all available measures, of both structural and non-structural nature.

All the assumptions mentioned above can be realized successfully only if local communities participate actively in the planning and consultation stage and then accept the whole strategy and implement it in practice. That participation is linked directly to social capital resources. Therefore, the next section presents those resources in Poland, as an important factor in controlling flood risk management.

SOCIAL CAPITAL RESOURCES IN POLAND

Resources of social capital in Poland, the density of social networks and the level of social trust are relatively low when compared to Western European countries (Czapiński, 2010). This is a heritage from the communist times when associational activity was strongly limited and people had to choose between family or close friendship-oriented networks, and state-controlled mass organizations. The notion of a "middle range social vacuum" was coined to describe the lack of independent civic associations during the communist era (Nowak, 1979), and this void has very slowly been filled only after 1989 (Gumkowska, Herbst, & Radecki, 2008; Kubiak & Miszalska, 2004). During these difficult times of the centrally planned economy many informal connections have been developed to deal with the shortages of goods (Paldam & Svendsen, 2000; Wedel, 2003). Nowadays, according to a public opinion survey (CBOS, 2010), 72% of adult persons in Poland do not participate in any non-governmental organisation activity, 15% declare that they spend some free time on voluntary activity in one thematic domain, and 5% are active in two different domains.

Although the level of social capital in Poland is relatively low, there are some important regional differences. They can be to some extent explained by historical processes which date back to the 19th century when Poland was partitioned between three neighbouring countries. Different economic and social structures of these states (legal systems, urbanization processes, development of the economy and infrastructure, culture, and mentality) led to different paths of development for more than one century (Bartkowski, 2003). Common stereotypical narratives identify the Prussian partition with order and thrift, the Austrian partition – with cultural autonomy, and the Russian partition – with backwardness and poverty (Jałowiecki & Szczepański, 2007). The territories that once were a part of the Russian Empire are thus assumed to have a lower level of social capital, both in terms of networks of associations and generalized trust. These lands are contrasted with the Austrian and Prussian partitions that are examples of areas with dense networks of cooperation (Zarycki, 2002). Additionally, Southern Poland (the Austrian partition) is the most fervently Catholic region in the country, and church organizations are an important part of the social capital landscape, with prevailing strong family and neighbourhood ties. The structures formed before 1918 were further modified with the changes in Poland's territory after the Second World War. They were followed by the mass transfer of population from former Polish lands, annexed to USSR, to former German lands, incorporated into Poland (the so-called Western and Northern Territories). Those areas are usually perceived as being characterized by low social activity on the one hand, and high population outflow, unemployment, and criminality on the other hand (Zukowski, 2003). However, Jałowiecki and Szczepański (2007) and Zarycki (2002) claim that these regions were mostly repopulated by younger and more entrepreneurial individuals who wanted to start a new life without strong social control, that is bonding social capital. The main outcome of this great social experiment was a new society, where bridging rather than bonding social capital should play a primary role.

The analysis of social capital resources in Poland shows their large spatial variability (Działek, 2009, 2011) (Fig. 2). They were measured at the level of statistical subregions (NUTS-3) on the basis of twelve indicators coming from various data sources (the national statistical office, databases of NGOs and church organisations, large social surveys). Indicators describe various aspects of social capital: density of non-governmental organizations; membership in arts, sports, and other hobby groups; membership in religious and church organisations; participation in local community activities and local public gathering; friendship networks; and level of social trust.

The principal component analysis was later conducted to identify two main dimensions of social capital – bonding and bridging. The component

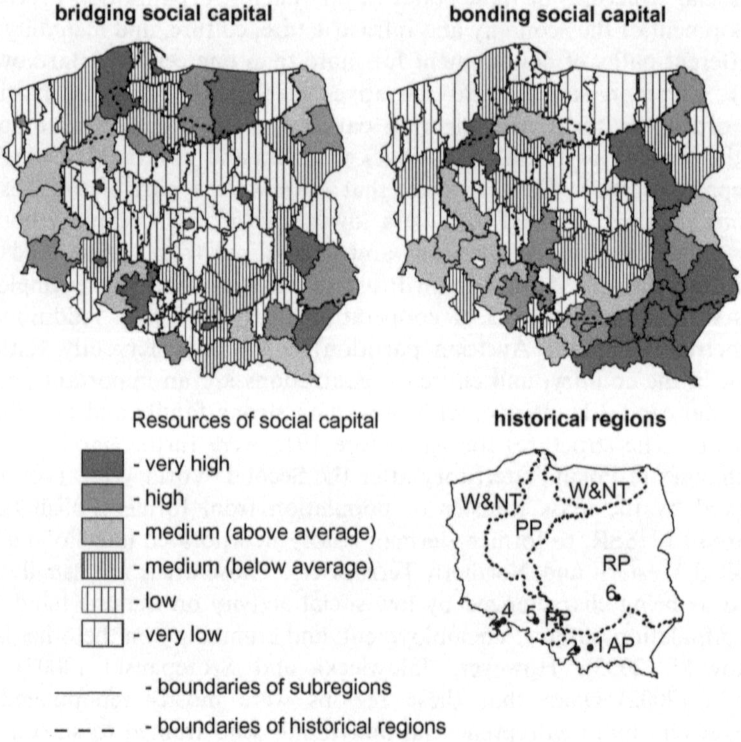

Fig. 2. Patterns of Social Capital Resources in Poland. Explanations: Statistical Subregions are Shown Only in the Two Upper Maps; Historical Regions: W&NT – Western and Northern Territories, PP – Former Prussian partition, RP – Former Russian partition, AP – Former Austrian Partition; Numbers Correspond to the Localities Included in the Study, see Table 1. *Source*: Data on social capital resources from Działek, 2011.

identified as bridging social capital includes variables describing density and membership of registered non-governmental organisations and the level of social trust. The second component describes bonding social capital and is represented by indicators of membership in religious organisations, local public participation, density of friendship networks and engagement in informal, not registered local organisations such as sport or arts groups.

The highest level of bridging social capital can be found in large cities, where many organisations of national and regional levels develop their activities. There are also many rural areas with high or at least medium resources of bridging social capital, mainly in the Northern and Western Territories and in south-eastern Poland (mostly the former Austrian partition). Those results show that in the Western and Northern Territories, where almost the whole population was affected by displacement and migration after the Second World War, the importance of associational activities is relatively high. The prevailing type of social organisation in that part of Poland resembles that typical for large cities which are also areas of immigration. On the contrary, in the areas that belonged to the Russian and Prussian partitions, the densities of civic networks and social trust are usually lower than average; in particular, the industrial area of Upper Silesia has the lowest resources of bridging social capital.

The spatial pattern of bonding social capital is significantly different from the pattern of bridging social capital. Its highest level is observed in Southern Poland (the entire Austrian partition) and in most areas of Eastern Poland. In turn, most of the Western and Northern Territories have very low resources of bonding social capital, with the exception of the Opole and Lower Silesia Regions. It is interesting to note that in large rural areas of Central Poland (from both the Russian and Prussian partitions), the community ties are also relatively weak, even though people living there are historically rooted in their region with much the same intensity as people living in Southern or Eastern Poland.

The results show that the civic activity, based mainly on a voluntary basis, is still rather low in Poland, although with some significant regional variation. Still, the conditions for implementing the flood risk management involving the local population might be consequently more difficult in Poland than in Western European countries. However, there are also regions with strong local bonds which might turn out to be an important asset for social capacity building. The management strategy has to involve most members of a certain community, not only those directly endangered but also those who are endangered only potentially and/or do not realise the danger. Therefore, active participation of the inhabitants' majority is essential and, at the same time, difficult to be achieved.

LOCAL RESPONSE TO THE FLOOD IN SOUTHERN POLAND

This section presents characteristic features of civic activity in a few selected communities in Southern Poland which experienced floods in the 1990s and 2000s. The results presented below make a part of the outcomes of a research project "Social attitudes and behaviour towards extreme phenomena", realized in the years 2004–2008 (Biernacki et al., 2008; Biernacki, Bokwa, Działek, & Padło, 2009). An overall analysis was conducted using data from nine localities, and answers for particular questions were shown for six selected localities, as they represent various historical regions and also differ in size, the number of inhabitants, and socio-economic character. The basic characteristics of the six localities chosen are given in Table 1, see also Fig. 1.

The research was based on questionnaires that were distributed in particular communities via local schools. In this chapter, we focus only on answers of 634 respondents, who admitted their households were at risk of flooding. The questionnaire survey was accompanied by in-depth interviews with the population at risk and focus group interviews with local leaders and risk managers. The schools selected for questionnaire distribution were located in areas where the inhabitants directly experienced large floods, according to the information from local and regional administrations. Therefore, we could obtain the opinions of the students' parents and grandparents living in those areas, as they were asked to answer the questions. The method used may be criticized for not providing a balanced demographic structure of the respondents, mainly with regard to the age structure. However, taking into consideration the large area of the study regions, specific criteria of respondents' selection (personal flood experience) and the assumption of having a large statistical population, the method seems to be the optimal solution. Application of a randomized mail survey was not possible due to high costs for such an undertaking. Also the application of telephone interviewing was limited by the fact that people directly affected by flood would make a small share of the whole population. The questionnaire consisted of questions concerning inhabitants' opinions, knowledge, attitudes, and behavior linked to flood

Table 1. Localities Used in the Study.

Locality	Settlement Type	No. of Inhabitants (Thousands) in 2006	Historical Region	Geographical Region	Flood Experience	Social Capital
1. Laskowa	Village	2.7	Former Austrian partition	The Beskid Wyspowy Mts., on the Lososina river	One large flash flood (1997)	High bonding social capital, medium bridging social capital
2. Maków Podhalański	Small town	5.7	Former Austrian partition	The Beskid Makowski Mts., on the Skawa river	One large flash flood (2001)	High bonding social capital, low bridging social capital
3. Polanica-Zdrój	Small town	7.0	Western & Northern Territories	The Stołowe Mts., on the Bystrzyca Dusznicka river	Two large flash floods (1997, 1998)	Low bonding social capital, medium bridging social capital
4. Kłodzko	Medium-sized town	28.1	Western & Northern Territories	Kłodzko Basin, on the Nysa Klodzka river	One large flash flood (1997), other floodings in 1998, 2006	Low bonding social capital, medium bridging social capital
5. Opole	Large city	127.6	Western & Northern Territories	Silesian Lowland, on the Oder river	One large slow flood (1997)	Low bonding social capital, high bridging social capital
6. Ostrowiec Świętokrzyski	Medium-sized town	73.7	Former Russian partition	Opatow Upland, on the Kamienna river	One large flash flood (2001)	Low bonding social capital, low bridging social capital

Source: Authors' own elaboration; social capital data based on Działek (2011).
Note: social capital resources presented in this table were calculated at the county (in Polish: *powiat*) level and they may in some cases differ from results shown in Fig. 2, where more general data at subregional level are presented.

phenomenon. Although the floods occurred four to eight years before the time of the survey in most of the cases, respondents' memories were still very vivid due to the severity of the flood events.

In order to show the links between social capital and actions undertaken, that is, the reactions resulting from the flood occurrence, the local civic response was studied. This is presented via the analysis of the respondents' answers to five questions. The data obtained were ordered according to the demographic features like age, sex, level of education, job, income, size of the locality, and the region it is located in.

The inhabitants declaring the individual response to the flood hazard make 18% of all respondents asked to answer the question: "Do you use any means to protect your house against a flood? If so, what is it?" For most demographic parameters, no differences were observed, with the exception of age and the locality size: older citizens undertake preventive actions more often than younger persons (11.7% among the age group 18–35 years, 19.6% among the respondents aged 36 years and more). The explanatory factor may be the previous experience of flood disaster. In the communities with high bonding capital, that is, small localities, the share was higher than in large cities – in villages about 45% of respondents were trying to protect their houses against floods, in small towns less than one third. In larger towns, less and less inhabitants decide to undertake protective actions (15% in medium-sized towns and 7% in large towns). This variability within communities of different sizes is overlaid by patterns resulting from historical processes: inhabitants from the area of the former Austrian partition, with high bonding social capital, undertake individual actions much more often (36%) than inhabitants from other regions (12%). The respondents were asked additionally about the purchase of insurance for their houses in case of a flood. A characteristic feature is the relatively low share of inhabitants taking that protective measure (below 50%). The differences between particular localities are connected with compulsory insurance for farmers, where both the agricultural land and the buildings must be insured. Fig. 3 shows the results for the six localities chosen against the mean values for the whole group and the differences described above are clearly seen.

The next question: "Have you undertaken together with other inhabitants any common actions in order to protect yourselves against a flood? If so, what is it?" refers to the collective response against floods and should reflect the general level of social capital. Less than one fifth of respondents declared undertaking collective mitigation behavior, which supports the thesis of generally low network capacities within Polish society.

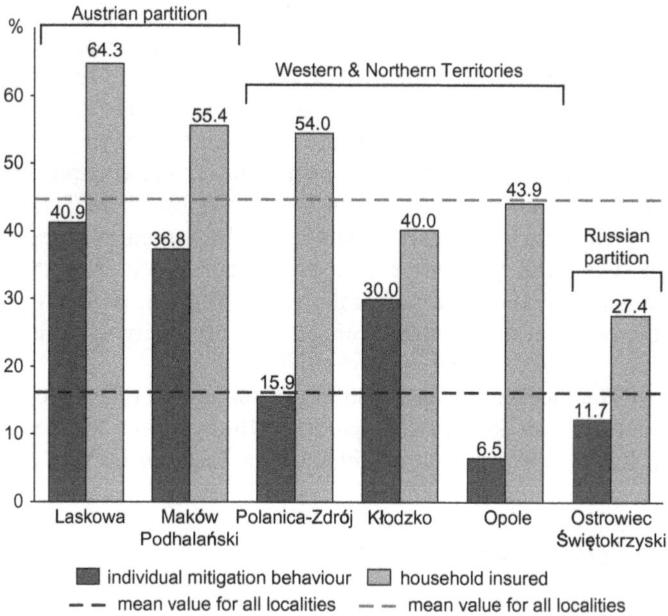

Fig. 3. Share of Respondents (%) Declaring Undertaking Individual Mitigation Behavior Against Floods and Purchasing Household Insurance in the Study Localities and in Historical Regions. *Source*: Authors' own Elaboration.

However there is some variability in this collective response, both among different demographic groups, as well as within territorial units. Common protective actions were declared more often by men (24%) than women (15.3%). Men declare undertaking common actions much more often than women because the works linked to flood prevention demand large efforts, usage of physical force, and special equipments; therefore some of them may be performed mainly by men. Citizens of age 36 and higher are declared to be more active (19.3%) than people of the age group 18–35 (14.2%). Inhabitants from villages (26%) and small towns (31.5%) were more active than the inhabitants from middle-sized towns (19.4%) and large cities (5.1%). That is related to the type of buildings and economic activity of the inhabitants in villages and large cities. People living in multistorey blocks of flats, not involved in agricultural activity, often neither see a need nor do they have the means for such cooperation.

Regional differences are also large: inhabitants from the area of the former Austrian partition undertake common actions much more often

(26.4%) than inhabitants from the former Russian partition (18.2%) or from Western & Northern Territories (11.1%). Data for particular localities are shown in Fig. 4. Historical traditions of civic activity in local communities in the former Austrian partition and their deep-rooting in the local environment may contribute to the high level of willingness to cooperate in case of flood. On the contrary, lower social capital resources in the former Russian partition and in Western & Northern Territories are connected with quite a low level of willingness for common actions in flood mitigation. Common preventive actions are correlated with the level of bonding social capital of a certain community. This can be seen in the comparison between rural and urban areas, and also between the historical regions.

A clear difference can be also seen in the perception of the local authorities accessibility, expressed in the answers to the question: "Have you alone or together with your neighbors contacted the local authorities concerning

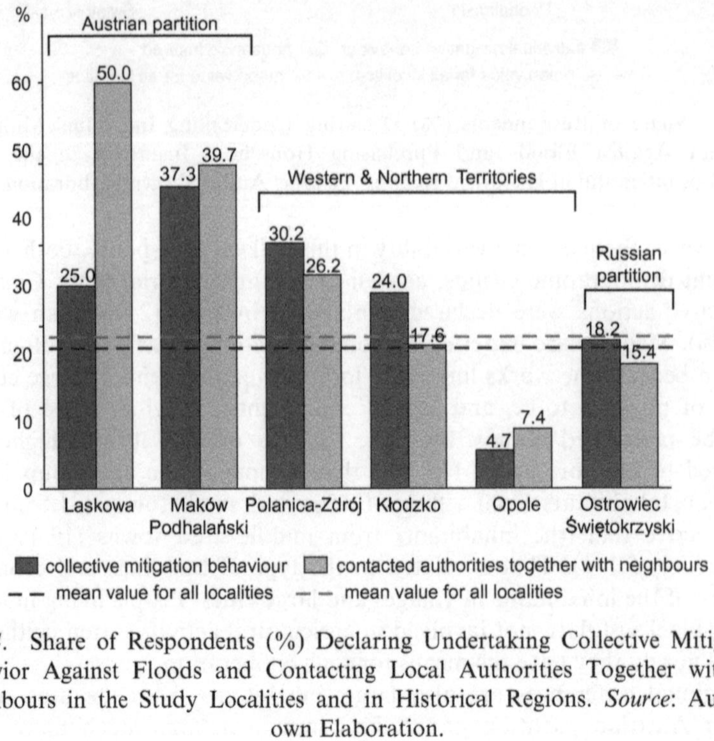

Fig. 4. Share of Respondents (%) Declaring Undertaking Collective Mitigation Behavior Against Floods and Contacting Local Authorities Together with the Neighbours in the Study Localities and in Historical Regions. *Source*: Authors' own Elaboration.

the protection of your houses against a flood?" Inhabitants from villages contact the representatives of local authorities much more often than the inhabitants in large cities, due to the necessity of such close cooperation in the everyday management of agricultural economy-oriented areas. Common social activities concerning cultural events, health protection etc., undertaken in villages, are reflected in the inhabitants' consciousness much more than similar actions organized in large cities where the amount of information reaching a person every day is much larger. Data for particular localities are shown in Fig. 4.

As many as 79% of all the respondents declared that the local authorities should be responsible for both organizing the preventive actions and undertaking all actions during the flood occurrence. The question was: "Who do you think should be mainly responsible for protecting your house against a flood? You and your family? You and your neighbours? Local authorities? Authorities of a higher administrative level?" A high share of respondents shows a very passive approach. There is usually no will to cooperate at the local level, and the trust in formal structures and procedures is lower than previously thought. Besides, those structures are designed to rather support and coordinate the citizens' activity but not to replace it totally. Also in that case the size of a locality had a significance in the share of answers declared. In the large cities 84.1% shifted the responsibility for any action to the local authorities, while in medium size towns it was 78.6%, in small towns 70.3%, and in villages 76.7%. As regards the historical background of the regions, the highest share was noted in Western & Northern Territories (82.4%), while in the former Austrian partition it was 76%, and in the former Russian partition it was 75.5%.

The inhabitants of the localities studied were also asked: "Have you thought about a permanent change of the place of residence and moving to a safer place where the flood hazard is much lower?" The results should be interpreted taking under consideration the historical factors and the variability of social bonds in particular regions. The communities from Western & Northern Territories and from the former Russian partition (Kłodzko, Opole, Ostrowiec Świętokrzyski) are less rooted in their regions and are more willing to move to other places/regions in order to avoid the flood destruction. An exception is Polanica-Zdrój which is a spa town; therefore, the inhabitants have strong economic bonds with that particular locality. In the former Austrian partition, the inhabitants are deeply-rooted in the local areas; they have strong emotional bonds with the land of their many-generation ancestors, and are deeply involved in the agricultural traditions

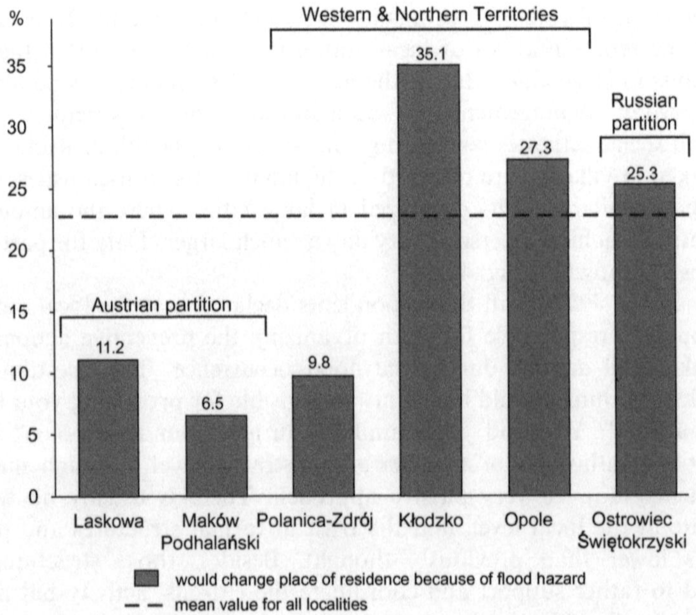

Fig. 5. Share of Respondents (%) Declaring the Will to Change the Place of Residence Because of Flood Hazard in the Study Localities and in Historical Regions. *Source*: Authors' own Elaboration.

of the region (Fig. 5). Additionally, a cognitive dissonance can play a significant role when the flood hazard is confronted with the affection for the place of living (Festinger, 1957).

CONCLUSIONS

The analysis presented above shows in general a high level of social passivity toward flood mitigation and preventive actions in Poland. However, clear spatial differences can be seen, connected with historical and social factors. The local response depends mainly on the size of the locality. That is combined with the different historical paths of social and economic development during the last two centuries, experienced by particular regions of Poland. Different regional values and norms towards collective behavior result in strengthening or weakening of local response to flood hazard. This is linked primarily to the bonding social capital; its high level is transformed into a high level of willingness to undertake common preventive actions. It is also related to a more frequent common engagement in contacting local authorities and emergency services. In localities with high bonding social capital, it can be expected that the larger part of the local community could get engaged in the flood risk management based on the social dialogue. In the communities with a high level of collective response, the individual response to flood hazard was also relatively higher. This can be a proof of the fact that social capital not only makes common actions easier to be organized but that it is also very useful in the information transfer concerning the flood hazard and preventive actions. Additionally, in those communities, which are deeply rooted in the local environment, there are more direct experiences of flood held in the community consciousness and memory and then transferred from the older persons to the younger ones. The resources of bonding social capital may be used in the risk communication strategies, where flood stories sharing can be used in building community resilience (McEwen & Jones, 2012).

The results obtained do not show a possibility to link the high level of bridging social capital with the local response to floods in Poland. Undertaking individual and common mitigation behavior in such communities, for example, with a high inflow of population in sub-urban areas or a high inflow of immigrants, is a real challenge for the flood risk management. Further research is necessary to see how weak ties and associational activity of the local population may be used in social capacities building and consequently strengthening of local response to flood.

The research presented above shows that social factors which influence the flood prevention activity are connected with long-term processes which result in cultural differences, characteristic of particular regions and localities of various size. As a consequence, there are significant difficulties in creating strategies aimed at the preparation of the citizens for a potential disaster. Factors which are usually considered as strongly linked with social activity, like education, income etc., directly connected with a particular person, seem to have less influence in that case. Social responses to floods turn out to be the consequence of wider historical and cultural factors which cannot be changed in a short time. Even the occurrence of a large catastrophic flood disaster, similar to the "millennium flood" in 1997, does not help to strengthen social bonds in a longer perspective, especially if those bonds have been weakened earlier and the local community has low resilience to any kind of social disruption.

NOTE

1. "Amoral familism" refers to a situation when strong bonding social capital is accompanied with weak bridging social capital. Communities with only strong social integration do not benefit from opportunities arising from weak ties (Granovetter, 1974), that is access to outside recourses. Their members "are discouraged from advancing economically, moving geographically, and engaging in amicable dispute resolution with outsiders" (Woolcock, 1998; p. 172).

ACKNOWLEDGEMENTS

The studies were conducted within a national research project titled: "Social attitudes and behavior towards extreme phenomena" which was organized and financed by the Ministry of Science and Higher Education of Poland during the period 2004–2008.

REFERENCES

Adger, W. N. (2003). Social capital, collective action, and adaptation to climate change. *Economic Geography*, *79*(4), 387–404.

Adger, W. N., Hughes, T. P., Folke, C., Carpenter, S. R., & Rockström, J. (2005). Social-ecological resilience to coastal disasters. *Science*, *309*(5737), 1036–1039.

Bartkowski, J. (2003). *Tradycja i polityka. Wpływ tradycji kulturowych polskich regionów na współczesne zachowania społeczne i polityczne* [Tradition and politics. Impact of culture traditions in the regions of Poland on contemporary social and political attitudes], Wydawnictwo Akademickie Zak [Zak Academic Publishing House], Warszawa.

Berke, P., Kartez, J., & Wenger, D. (1993). Recovery after disaster: Achieving sustainable development, mitigation and equity. *Disasters*, *17*, 93–109.

Biernacki, W., Bokwa, A., Domański, B., Działek, J., Janas, K., & Padło, T. (2008). Mass media as a source of information about extreme natural phenomena in Southern Poland, [in:] A. Carvalho (red.), Communicating Climate Change: Discourses, Mediations and Perceptions, Centro de Estudos de Comunicação e Sociedade, Universidade do Minho, Braga, 190–200. Retrieved from http://www.lasics.uminho.pt/ojs/index.php/climate_change/

Biernacki, W., Bokwa, A., Działek, J., & Padło, T. (2009). Społeczności lokalne wobec zagrożeń przyrodniczych i klęsk żywiołowych [Local communities and natural hazards and disasters], IGiGP UJ [Publishing Unit of the Institute of Geography and Spatial Management, Jagiellonian University], Kraków.

Bolin, R., & Stanford, L. (1998). The Northridge earthquake: Community-based approaches to unmet recovery needs. *Disasters*, *22*, 21–38.

Bourdieu, P. (1986). The forms of capital. In J. G. Richardson (Ed.), *Handbook of theory and research for the sociology of education* (pp. 241–258). New York, NY: Greenwood Press.

CBOS. (2010). *Aktualne problemy i wydarzenia* [Current problems and events], 236, CBOS [Public Opinion Research Center], Warszawa.

Coleman, J. S. (1994). *Foundations of social theory*. Cambridge: First Harvard University Press.

Cutter, S. L., Boruff, B. J., & Shirley, W. L. (2003). Social vulnerability to environmental hazards. *Social Science Quarterly*, *84*(2), 242–161.

Czapiński, J. (2010). Social capital. In J. Czapiński & T. Panek (Eds.), *Social diagnosis 2009. The subjective quality and objective conditions of life in Poland* (pp. 141–150). Warsaw: The Council for Social Monitoring.

Dynes, R. (2002). *The importance of social capital in disaster response*, Preliminary Paper #327. University of Delaware, Disaster Research Centre.

Działek, J. (2009). *Social capital and economic growth in Polish regions*. MPRA working paper. Retrieved from http://mpra.ub.uni-muenchen.de/18287/1/MPRA_paper_18287.pdf. Accessed on February 26, 2013.

Działek, J. (2011). *Kapitał społeczny jako czynnik rozwoju gospodarczego w skali regionalnej i lokalnej w Polsce* [Social capital as the economic development factor in regional and local scale in Poland], Wydawnictwo Uniwersytetu Jagiellońskiego [Jagiellonian University Publishing House], Kraków.

Elliot, J. R., Haney, T. J., & Sams-Abiodum, P. (2010). Limits to social capital: Comparing network assistance in two New Orleans neighborhoods Devastated by Hurricane Katrina. *The Sociological Quarterly*, *51*, 624–648.

European Commission. (2003). *Best practices on flood prevention, protection and mitigation*. Retrieved from http://ec.europa.eu/environment/water/flood_risk/pdf/flooding_bestpractice.pdf. Accessed on March 10, 2013.

European Commission. (2004). *Communication from the Commission to the Council, the European Parliament, the European Economic and Social Committee and the Committee of the Regions*. Commission of the European Communities, Brussels.

Festinger, L. (1957). *A theory of cognitive dissonance*. Stanford, CA: Stanford University Press.

Gittel, R. J., & Vidal, A. (1998). *Community organizing: Building social capital as a development strategy*. Thousand Oaks, CA: Sage.

Granovetter, M. S. (1974). *Getting a job: A study of contacts and careers*. Cambridge, MA: Harvard University Press.

Gumkowska, M., Herbst, J., & Radecki, P. (2008). *Podstawowe fakty o organizacjach pozarzadowych. Raport z badania 2008* [Basic facts about non-govermental organisations. Report of 2008], Stowarzyszenie Klon/Jawor [Klon/Jawor Association], Warszawa.

Hyyppä, M. T. (2010). *Healthy ties: Social capital, population health and survival*. New York, NY: Springer.

Instytut Meteorologii i Gospodarki Wodnej, Instytut Rozwoju Miast, Instytut Ochrony Środowiska, Kraków [Institute of Meteorology and Water Management, Institute of Urban Development, Institute of Environmental Protection] (2011). *Prognoza oddziaływania na środowisko skutków realizacji program ochrony przed powodzia w dorzeczu Górnej Wisły* [Prognosis of the environmental effects of the flood protection programme realisation in the Upper Vistula River Basin]. Retrieved from http://www.malopolska.uw.gov.pl/doc/Prognoza_oddzialywania_gorna_Wisla__kwiecien_2011.pdf. Accessed on March 10, 2013.

Jałowiecki, B., & Szczepański, M. S. (2007). Dziedzictwo polskich regionów [Heritage of Polish regions]. In A. Kojder (Ed.), *Jedna Polska. Dawne i nowe zróżnicowania społeczne* [One Poland. Former and new social differences], Wydawnictwo WAM, Polska Akademia Nauk [WAM Publishing House, Polish Academy of Sciences], Kraków, 59–87.

Kaniasty, K., & Norris, F. H. (1995). In search of altruistic community: Patterns of social support mobilization following Hurricane Hugo. *American Journal of Community Psychology, 23*, 447–477.

Kubiak, A., & Miszalska, A. (2004). Czy nowa próżnia społeczna, czyli o stanie więzi społecznych w III Rzeczypospolitej [New social vacuum? State of social bonds in the Third Republic of Poland]. *Kultura i Społeczeństwo, XLVIII*(2), 9–43.

Kuhlicke, C., Steinführer, A., Begg, C., Bianchizza, C., Bründl, M., Buchecker, M., De Marchi, B., ... Faulkner, H. (2011). Perspectives on social capacity building for natural hazards: Outlining an emerging field of research and practice in Europe. *Environmental Science and Policy, 14*, 804–814.

Kuhlicke, C, Steinführer, A., Begg, C., & Luther, L. (2012). *Toward more resilient societies in the field of natural hazards: CapHaz-Net's lessons learnt. CapHaz-Net WP10 Final Report*, Helmholtz Centre for Environmental Research – UFZ: Leipzig & Johann Heinrich von Thünen Institute: Braunschweig, 2012. Retrieved from http://caphaz-net.org/outcomes-results/CapHazNet_Del%2010.3_final.pdf. Accessed on May 14, 2013.

Maciejewski, M. (2000). *Model kompleksowej ochrony przed powodzia w obszarze dorzecza górnej Wisły na przykładzie województwa małopolskiego* [Model of complex flood protection in the Upper Vistula River Basin with the example of the Malopolska region], OKI RZGW [Regional Office for Water Management], Kraków.

McEwen, L., & Jones, O. (2012). Building local/lay flood knowledges into community flood resilience planning after the July 2007 floods, Gloucestershire, UK, *Hydrology Research*. doi:10.2166/nh.2012.022

Mihaylova, D. (2004). *Social capital in Central and Eastern Europe. A critical assessment and literature review*. Budapest: Center for Policy Studies, Central European University.

Murphy, B. L. (2007). Locating social capital in resilient community-level emergency management. *Natural Hazards, 41*(2), 297–315.

Nowak, S. (1979). System wartości społeczeństwa polskiego [The system of values in Polish society]. *Studia Socjologiczne, 4*, 155–172.

Paldam, M., & Svendsen, G. T. (2000). An essay on social capital: Looking for the fire behind the smoke. *European Journal of Political Economy, 16*, 339–366.

Paxton, P. (2002). Social capital and democracy: An interdependent relationship. *American Sociological Review, 67*(2), 254–277.

Pelling, M. (1998). Participation, social capital and vulnerability to urban flooding in Guyana. *Journal of International Development, 10*, 469–486.

Portes, A. (1998). Social capital. Its origins and applications in modern sociology. *Annual Review of Sociology, 24*, 1–24.

Portes, A. (2000). The two meanings of social capital. *Sociological Forum, 15*(1), 1–12.

Putnam, R. D. (2000). *Bowling alone: The collapse and revival of American community*. New York, NY: Simon & Schuster.

Putnam, R. D., Leonardi, R., Nanetti, R. (1993). *Making democracy work*. Princeton, NJ: Princeton University Press.

Sander, T., & Lowney, K. (2006). *Social capital building toolkit version 1.2*, Saguaro Seminar: Civic Engagement in America, John F. Kennedy School of Government, Harvard University.

Starkel, L., (Ed.). (1999). *Geografia Polski: środowisko przyrodnicze* [Geography of Poland: natural environment]. PWN, Warszawa.

van Deth, J. W. (2003). Measuring social capital: Orthodoxies and continuing controversies. *International Journal of Social Research Methodology, 6*(1), 79–92.

Wedel, J. R. (2003). Dirty togetherness: Institutional nomads, networks, and the state–private interface in Central and Eastern Europe and the former Soviet Union. *Polish Sociological Review*, *2*, 139–159.

Westlund, H., & Adam, F. (2010). Social capital and economic performance: A meta-analysis of 65 studies. *European Planning Studies*, *18*(6), 893–919.

Woolcock, M. (1998). Social capital and economic development: Toward a theoretical synthesis and policy framework. *Theory and Society*, *27*, 151–208.

Yohe, G., & Tol, R. S. J. (2002). Indicators for social and economic coping capacity: Moving toward a working definition of adaptive capacity. *Global Environmental Change*, *12*, 25–40.

Zarycki, T. (2002). *Region jako kontekst zachowań politycznych*. Wydawnictwo Scholar, Warszawa.

Zukowski, T. (2003). Wybory samorzadowe 2002 i referendum akcesyjne 2003 – perspektywa subregionów [Local elections of 2002 and EU accession referendum of 2003 – subregional perspective]. In M. Kowalski (Ed.), *Przestrzeń wyborcza Polski* [Voting patterns in Poland], PAN-IGiPZ [Polish Academy of Science], Warszawa, 29–48.

CHAPTER 9

ENHANCING LOCAL RESPONSES THROUGH DISASTER RESILIENCE IN SCHOOLS AND COMMUNITIES IN JAPAN

Rajib Shaw and Shohei Matsuura

ABSTRACT

Schools play an important role in Japan by becoming evacuation centers after disasters. Depending on the nature of disaster, the school can be occupied for several days to several months. Therefore, schools play a crucial role in disaster risk reduction and can contribute to very strong bonding with the local communities. This chapter describes the experiences of six cities with the roles of schools during disasters. Kamaishi, Kesennuma, and Natori, three cities affected by the tsunami, have shown the important role that schools played in the time of disaster. Although some schools were destroyed in these three cities, people spent significant time in other schools as evacuees. Pre-disaster preparedness of schools and communities helped a lot in this regard. Taking the experiences from the East Japan disaster, Saijo, Owase, and Oobu cities in West Japan demonstrated their preparedness for future disaster. The chapter also shows that school-centered disaster preparedness before the disaster

Risks and Conflicts: Local Responses to Natural Disasters
Community, Environment and Disaster Risk Management, Volume 14, 207–230

ISSN: 2040-7262/doi:10.1108/S2040-7262(2013)0000014015

leads to an effective role during the disaster and also facilitates post-disaster recovery with schools as the center.

Keywords: School community linkages; school centered recovery; East Japan Earthquake and Tsunami; local response; disaster preparedness

INTRODUCTION

According to the Ministry of Education, Culture, Sports, Science and Technology (MEXT) of the Government of Japan, there were about 42,000 public schools in 2009. The Japanese public schools include kindergarten, elementary school, junior high school, high school, and special-needs school. The public school has two roles –as a place of learning and living for children and as a core of the community.

In its role as a place of learning and living for children, the Central Council for Education of MEXT describes schools as follows: A school should provide balanced education for the attainment of knowledge and moral and physical health during the developmental stage of children. In addition, schools should contribute to lifelong learning. In particular, focusing on mental development are important to enhance the academic ability and make a base of learning for life. Also, schools are fundamental in developing a good heart in terms of humanity and social relations through communal living with friends of the same generation. Most important, it is essential to discover the strengths of each child as well as to enhance their character and ability. Thus, every public school has course instruction and daily life guidance based on the proposal of the MEXT (Suda, 2012).

Schools are now addressing disaster education in addition to course instruction and daily life guidance. The importance of disaster education at the school level is recognized in the works of Radu (1993), Kuroiwa (1993), Arya (1993), Frew (2002), and Shaw, Shiwaku, Kobayashi, and Kobayashi (2004; Shaw, Shiwaku, & Takeuchi, 2011). Also, Shiwaku, Fujieda, Takeuchi, and Shaw (2010) report that disaster education in school is effective in raising awareness not only of students but also of their family members and the community. A school plays an important role in raising awareness among students, teacher, and parents (Shaw & Kobayashi, 2001). On the other hand, Okada (2008) states that schools are important places as they are the core of the community and serve as evacuation sites in times of disaster. For example, not only children but also parents participate in many events in schools, such as athletic and cultural festivals and class observation day. In this way, exchanges among the participants are deepened. Schools are also used as polling places during elections. Lastly, schools are used as evacuation sites, when flood or sediment disasters caused by heavy rain occur or when a big disaster happens.

With this context, this chapter analyzes the roles of school as evacuation centers and their effectiveness in linking to the community. Citing six examples from tsunami-affected East Japan and future high-risk West Japan, the chapter reviews the current level of preparedness in these cities and then draws some common lessons that can be applied elsewhere as a local response through strengthening links between schools and communities.

SCHOOLS AS AN EVACUATION SITE

As explained in the previous section, schools are used as evacuation sites. According to the survey of the Fire and Disaster Management Agency (FDMA) in 2008, schools account for 60% of the public facilities used as disaster prevention facilities (NIER, 2008). Furthermore, according to NIER, in 2011, 89.3% of all public schools in Japan were designated as evacuation sites. Also, municipal schools account for 91.8% of the public schools used as evacuation sites. Most of the elementary and junior high schools are administered by the municipalites in Japan. Hence, public elementary and junior high schools are the schools primarily used as evacuation sites.

There are three reasons why schools are often used as evacuation sites (CAO, 2011). First, because it is stipulated by the Disaster Law; the second is because of schools' potential as infrastructure; and the third is the high intimacy and visibility of schools. The Disaster Countermeasures Basic Act, which was enacted in 1961, states that municipalities have a big responsibility for disaster countermeasures. Therefore, the municipality sets up public facilities as evacuation sites when it is likely that a disaster will happen or when a big disaster does happen. Sixty percent of the public facilities in municipalities are elementary and junior high schools. Therefore, it is often the case that elementary schools and junior high schools are used as evacuation sites. A second reason why schools are frequently used as evacuation sites over other public facilities, such as community centers, is the high potential of school facilities. Schools have huge grounds, which are useful as temporary evacuation and waiting sites. Also, schools can be used as parking places, collection sites of relief supplies, and places for evacuation tents, temporary baths, and command posts. In particular, the gymnasium can be used as shelter because it has a large indoor space and is strongly constructed. In addition, there are many toilets and washrooms. For these reasons, the school facility has a high potential as evacuation site over other public facilities. Finally, because everyone in the community has gone to the municipality's elementary school and junior high school in the past, the school is a place where close and intimate relations are formed, and schools are more highly visible than other public facilities.

Thus, school functions not only as a place of education for children and students but also as a core of the community and disaster prevention facility.

UTILIZATION OF SCHOOLS IN PAST DISASTERS

In this section, three examples of disasters are shown – the Great Hanshin-Awaji Earthquake as an example of an earthquake in an urban area; the Mid Niigata Prefecture Earthquake as an example of an earthquake in a mountainous area; and the Great East Japan Earthquake as an example of an earthquake and tsunami in a huge area (Suda, 2012). These are examples of three big disasters in Japan in which the damage was different according to the region of occurrence (Table 1).

The Great Hanshin-Awaji Earthquake happened in an urban area in 1995. Infrastructures such as water and sewerage, electricity, gas, and telephone suffered considerably. The restoration of electricity took about 1 week, gas about 3 weeks, and water about 3 months. Rescue work, firefighting, and transportation of materials became difficult because the roads and harbor also suffered damage. Because of this, the evacuated people reached around 310,000 at its peak and about 1,100 evacuation sites were set up, among which were 390 public elementary and junior high schools. They took in around 180,000 evacuees (NIER, 2008). Subsequently, about 60% of the people evacuated used schools as evacuation sites. In Kobe, 98% of public schools in Higashinada-ku, Nada-ku, Cyuo-ku, Hyogo-ku, Nagata-ku, and Suma-ku were used as evacuation sites.

The Mid Niigata Earthquake happened in a mountainous area in 2004. Because of the earthquake, several sediment disasters and road destruction

Table 1. Utilization of Schools as Evacuation Site in Disasters.

	Great Hanshin-Awaji Earthquake	Mid Niigata Prefecture Earthquake	Great East Japan Earthquake
Date	January 17, 1995	October 23, 2004	March 11, 2011
Magnitude	7.2	6.8	9.0
Intensity of Japan Meteorological Agency	7	7	7
Affected area	Urban area (metropolis)	Country area (mountain)	Urban area (local city) Rural area (coastal)
Human suffering	Dead or missing: 6,436 Injured: 43,792	Dead: 59 Injured: 4,795	Dead: 15,835 Missing: 3,669 National Police Agency (NPA), November 7, 2011
Building damage	Complete collapse: 104,906 Half collapse: 144,274	Complete collapse: 3,175 Half collapse: 11,609	Complete collapse: 120,204 Half collapse: 189,164
Number of people evacuated	316,678 (according to the Cabinet Office)	103,178 (according to the Cabinet Office)	386,739 (according to the Cabinet Office) 1 week later
Number of evacuation sites	About 1,100	About 600	101,640 3 months later
Proportion of schools used as evacuation place (at peak)	About 35%	About 30%	About 30% 1 week later
Proportion of people who used the schools	About 60%	About 50%	About 40% 3 months later

occurred. As a result, 61 settlements became isolated in the cities of Nagaoka and Ojiya and in the village of Yamakoshi. In terms of lifelines, electricity was cut off in about 300,000 houses, 130,000 houses were without water, and 50,000 houses were without gas immediately after the disaster. Several areas spent 2–3 months to connect with the rest of the country because of damaged roads. Because the area damaged was wide, the number of evacuees reached about 100,000 at its peak. According to a survey of Niigata Prefecture, 50% of the people were evacuated to schools 5 days after the disaster. They used the gymnasiums, several classrooms, and the grounds.

The Great East Japan Earthquake happened in 2011. From the Iwate Prefecture to the Fukushima Prefecture, the tsunami brought extensive damage. As a result, many people lost their houses and around 380,000 people were evacuated. According to a survey of the Cabinet Office in June 2011 (3 months after the disaster), schools accounted for 30% of all evacuation sites and 40% of the evacuated people used schools as evacuation sites.

In each of the three cases, 40–50% of the people used the public school as evacuation site, emphasizing the importance of schools as facilities to protect people during a disaster. School is also used as evacuation site during flood and sediment disasters. The occurrence of flood and sediment disasters can be predicted by means of precipitation amounts. Therefore, when risks are identified by the Japan Meteorological Agency (JMA) and the local government, evacuation instructions or directives are announced by the local government. In this case, people evacuate to the nearest evacuation site, such as a school. In fact, schools are used not only after disasters but also before disasters.

As an example, Japan suffered a great deal of damage from Typhoon No. 2 and No. 12 in 2011. In many parts of Japan, record rainfall was observed. As a result, flood and sediment disasters happened in many parts of Japan. Before the flood and sediment disasters happened, people moved to evacuation sites because an evacuation directive was announced. After the flood and sediment disaster happened, some people continued to stay in the evacuation sites because their houses had been damaged or the residence area still at risk for disasters. As an example of flood disaster risk management, in the city of Shiso in Hyogo Prefecture, people evacuated to the public high school because the water level of river crossed over the danger level.

As discussed above, schools have functioned as evacuation site in disasters, including earthquakes, tsunamis, and flood and sediment disasters.

Massive earthquakes and tsunamis such as the Great Hanshin-Awaji Earthquake and the Great East Japan Earthquake might occur infrequently, but heavy rain and precipitation shows a pattern of increase in recent years, indicating that schools may be used as evacuation sites more frequently. Therefore, it is important to make arrangements to take in evacuated people in schools anytime. For this reason, links between schools and their communities become very important.

CASES OF CITIES AFFECTED BY EAST JAPAN EARTHQUAKE TSUNAMI

Based on the above discussion, here we will describe damages and responses of school and communities in three cities: Kamaishi, Kesennuma, and Natori in the affected region of the Tohoku disaster (East Japan Earthquake and Tsunami (EJET)).

Kamaishi Experiences

EJET brought the following damages to Kamaishi City: 1,046 deaths and missing persons (as of March 2012) and an estimated 10,266 evacuees. Damage occurred to 4,614 houses, 4 kindergarten facilities (1 completely destroyed), 9 elementary schools (2 completely destroyed, 7 partially destroyed), and 5 junior high schools (2 completely destroyed, 3 partially destroyed). In the Unosumai district, the tsunami reached the third floor of a building. Currently, 3,200 temporary housing units are still in use in the city.

Resumption of schooling faced difficulties because of the damage to both the elementary and junior high schools. However, the city Board of Education (BoE) arranged for the students of damaged schools to attend other schools. In addition, BoE made decisions not to construct temporary housing facilities on the school grounds so that students will be able to use them for sports and other activities. Currently, children are bussed to school.

More than 1,000 people were reported dead or missing in the city, including 8 students and 5 teachers, as the result of the EJET. Although the school DRR program in Kamaishi City proved to be effective up to a certain point, the voluntary evacuation of schoolchildren during EJET is not considered a success of the DRR program because some students did become victims. Therefore, the city aims to further strengthen the local DRR system to make use of the lessons learned from the EJET. Teachers must share DRR education materials with each other, which are based on the Guidelines for Tsunami DRR Education. Existing documents and manuals will be revised and incorporated with global DRR trends to maintain community awareness of the DRR system.

Previous DRR efforts included cooperation between schools and local communities, such as identification of evacuation shelters with parents on class visit days. In addition, schools, in cooperation with town associations and "School Guards" consisting of community members, have assured schoolchildren and parents that children will be evacuated to safety when disaster occurs and family members or parents will come to pick them up after the disaster. Evacuation drills have been performed, and handover training has been offered to parents after school hours (handover of students to parents until safety has been confirmed is prohibited). The BoE has been encouraging parents to have confidence that their children will be evacuated to safety. Junior high school students have learned about disaster response and volunteer opportunities in their ethics classes. They also receive practical DRR education in cooperation with the local authorities and participate in DRR volunteer activities.

A disaster management system has been developed in the city of Kamaishi for all schoolchildren and their parents, based on the themes below. Together with the local authorities, all parties must cooperate in order to protect the safety of children in the event of a disaster.

Theme 1: Improvement of DRR technology–related teaching materials
Theme 2: Development and enforcement of the training program for schoolteachers
Theme 3: Development and implementation of a practical DRR education program
Theme 4: Implementation of innovative initiatives in disaster response with local contexts

Using these themes, Kamaishi has fostered a DRR culture that states, "If you live in Kamaishi, you must prepare for tsunamis" and has aimed for citizens to develop skills to protect themselves and help each other.

The BOE of Kamaishi aims to teach respect for other people and all forms of life by reinforcing safety education (traffic rules, safe practices in daily life, and disaster response) and, at the same time, eradicate issues such as bullying. This safety education and associated school security plan along with a crisis control manual have been developed in cooperation with relevant organizations. Furthermore, schools and local governments make sure that people do not get satisfied just by developing plans and manuals, but to look into the practicality of how these will be utilized in the most effective way and by whom.

Kesennuma Experiences

The EJET brought the following damages to the city of Kesennuma: (1) There were about 1,400 victims (include 260 missing), as well as losses of livelihoods and property (e.g., houses, shops, cars) and losses of economic (e.g., markets, factories, fields) and educational infrastructures (e.g., schools, community centers, libraries). (2) In affected areas, community ties were lost because of people being displaced to shelters and temporary housing. (3) The tsunami damaged and shut down lifelines – communication networks and transportation networks (road and railways). In this situation, school officials were forced to respond using their own judgment in the absence of information and without contact with BoE and other related local government offices.

With aims to resume school activities in a disaster situation, BoE and schools were faced with many challenges, such as confirmation with regard to the extent of damage, locations of students, repairing facilities, and restoring lifelines. As of now, school bus service is available for transporting students residing in temporary housing over a wide area. Early provision of nutritious school meals was established, despite the damages and economic difficulty. Workshops for physical and psychiatric care have also been offered for children and teachers affected by the disaster. Although still not up to their previous capacity, Kesennuma was able to resume classes for all elementary and junior high school students on April 21, 2011, stating, "We will start with what we have. At least we have classrooms and food." BoE also adjusted the number of school days and hours (e.g., by shortening summer vacation) to minimize inequalities in learning hours, especially for students who had to transfer to another school. Playgrounds and gymnasiums have been temporary rehabilitated, and school and BoE events have been carefully selected and organized to ensure continuity of education for students in the region.

The schools, which were isolated after the disaster, were tasked with taking urgent measures to protect the children and the local residents. Immediately after the earthquake, the schools faced difficulties in the handover of students to their parents in the midst of first- and second-stage evacuations, when the schools did not have sufficient stockpiles or accommodation facilities and students had to take shelter without blankets or warm clothing despite the cold weather. Schools had difficulty operating the shelters and managing the schools under these severe conditions. Other challenges included lack of supplies and fuel, confirming safety of students and reporting to BoE, receiving and distributing emergency supplies (water, food, and clothing). These were soon followed by problems making decisions for school management and rescheduling postponed activities and school events (e.g., graduation ceremonies), arranging transportation and scheduling for schoolteachers (e.g., arranging night duty), and confirming the safety of school officials and their families. Challenges were also identified, including the timing of supervisory shifts from shelter to school management of the buildings and creating a system for providing psychological care to teachers affected by disasters.

In the current recovery and reconstruction process, schools are working to resume normal school operations and, at the same time, to provide support for people in shelters and temporary housing. Future efforts will aim to respond according to individual circumstances of the schools (e.g., location and extent of damage). In addition, because the time that children spend at school is not a large percentage of their everyday lives, children must be trained to use their own judgment when faced with disasters, especially on their way to and from school and at home. DRR education that builds capacity to respond to crises and facilitate coordination between communities and schools by encouraging the whole community to look over and protect their children and promoting mutual help together with town associations and volunteers is important to cope with future disasters.

To improve DRR education, incorporating the lessons from EJET and creating a system that will protect the lives of children both inside and outside schools, revised plans and institutional arrangements need to be developed in cooperation with schools, town associations, city governments, and specialized agencies. Moreover, it is necessary to overcome bias with regard to normalcy and majority of people have dependency during emergencies. Further strengthening of DRR should be aimed with the concept of "N-help" (nonprofit organizations (NPOs), nongovernmental organizations, network, new) in addition to self-help, mutual help, and public help. In the next steps, enhancing the role of schools to become DRR hubs and disseminating records and lessons learned at schools from the EJET experience are planned.

Natori Experiences

The EJET brought following damages to the city of Natori. In Natori, 911 people died and 42 were reported missing. More than 5,000 buildings were completely or mostly destroyed. Because graduation ceremonies (which ended in the morning) were held in five junior high schools across the city on the day of EJET, 17 junior high school students lost their lives. Just after the earthquake, electricity to the entire city was cut off and telephone communications were suspended. Gas, water supplies, and sewage systems were also extensively damaged. In effect, the city was paralyzed. Although electricity was restored relatively quickly (about 4 days after the EJET), nearly 1 month passed before the lifelines were restored (longer in tsunami-damaged areas). Damage caused by the earthquake was minimal considering its size; however, the damage due to the tsunami was enormous. All deaths and missing people can be attributed to the tsunami. The majority of housing in the coastal areas, which consisted largely of wooden houses within 1 km of the sea, was washed away; as was the case in the Yuriage district, in which fires also occurred. Coastal areas of the Yuriage and Shimomasuda districts suffered catastrophic damage from the tsunami. Many residents in the Yuriage district thought that a big tsunami would not come because the coastline in the area is flat. In addition, because no damage was reported in the district from the previous earthquake in Chile, the residents doubted the accuracy of the early warnings, delaying evacuation during the EJET. Another factor was the time lag between the earthquake and the tsunami, during which many residents waited, leaving them insufficient time to escape when the tsunami actually arrived. Yuriage, in particular, was strongly affected because there are no hills surrounding the district. Insufficient storage of blankets, water, and other emergency supplies exacerbated the situation. Toilet facilities and other sanitary factors after the tsunami were also a problem. All these factors were identified as important lessons from the EJET. After the EJET, schools were faced with providing psychiatric care for children and teachers. Education of the children from the damaged schools resumed in available classrooms in other schools.

Assuming future disasters will occur, the psychiatric care of students, teachers, and residents in affected areas must be considered during and after disasters. In addition, although many schools are now earthquake-proofed, schools are still insufficient to function as DRR facilities. And the handover of students to their parents and management of evacuation shelters remain to be improved in the future.

Currently, considerations for new schools include rebuilding on raised land (3–5 m in height) in order to improve schools acting as evacuation centers and as DRR hubs. Concrete plans to engage local communities in the recovery process and disaster-proofing of schools are being developed. However, community and school recovery has been a slow process so far.

In the future, improvements in DRR education in Natori will further emphasize that communities should protect themselves, and in this context, the new DRR manual is being developed. As children do not spend much time in school, the ability to save their own lives during disasters is also important outside of schools. In addition, comprehensive DRR measures, which include not only tsunami, but also other disasters, must be implemented with considerations of risk management and disaster probabilities to conduct practical disaster prevention drills.

CASES OF CITIES PREPARING FOR FUTURE DISASTERS

As discussed above, the EJET caused severe damages. Because of preparedness and school–community linkages, the schools played a vital role during and after the disaster. There are several cities that are now preparing for the next big disaster, which may happen in western Japan, known as Nankai, and Tonankai earthquake and tsunami. This section describes experiences from three cities: Saijo (Ehime Prefecture), Owase (Mie Prefecture), and Oobu (Aichi Prefecture). All these cities are at high risk for future earthquakes and tsunamis.

Experiences of Saijo

The city of Saijo, in the Ehime Prefecture, became a city with population of 116,455 and a land area of 509 km^2 after three towns were newly incorporated to the city in 2004. The geography can be characterized as having a narrow plain area (comprising 30% of the total), sandwiched by a mountain range and the shoreline. In 2004, six typhoons that landed in Shikoku Island triggered record rainfalls, causing damage to 2,774 public facilities and 29 deaths in the prefecture, of which 5 deaths were in Saijo. In Saijo, there are three tectonic lines, and the median tectonic line runs through the southern part of the city. For this reason the city is designated by the Act on Special Measures Concerning Advancement of Countermeasures against Disasters of Tonankai and Nankai Earthquakes as a Region for the Advancement of Countermeasures against Disasters of Tonankai and Nankai Earthquakes.

After the typhoon disaster experience in 2004 and from the predictions of 70 and 60% probability of earthquakes occurring in Tonankai and Nankai, respectively, within the next 30 years, Saijo has revised its DRR planning with the following four concepts with guidance from Graduate School of Global Environmental Studies at Kyoto University.

- Make citizens the main actors of DRR: Make every citizen aware that they are the main actors in DRR
- Build mechanism to save the most vulnerable people: Build mechanism so that no vulnerable group exists in the city
- Share regional DRR culture: Share DRR culture throughout the city extensively
- Deliver message to next generation to build a disaster-resilient society: Train young leaders with DRR capacity

With the above, Saijo has started implementation of the 12-Year-Old Education Project from 2006. The project aims to build a community-centered DRR culture by training young leaders with social skills and DRR capacity through acquirement of DRR knowledge. Although 12-year-olds (6th-graders) are in the vulnerable group, they are capable of thinking and making decisions by themselves; therefore, this group is seen as a suitable target for starting DRR education. For implementation, the Working Committee (consisting of the faculty of each school) in coordination with the Disaster Management Division of the city government and with Ministry of Land, Infrastructure, Transport, and Tourism (MLIT), the fire department, Nippon Telegraph and Telephone Corporation (NTT), and disaster managers on an as-needed basis will implement the project (Fig. 1).

The main annual events of the project are the leader's training and the DRR Summit. For the DRR Summit, about 60 sixth-grade students from 26 elementary schools decide what DRR topics they would like to study. In the DRR camp (leaders' training), conducted during summer vacation, the students acquire DRR knowledge and skills and conduct such activities as earthquake simulation vehicle exercise, first-aid class, exercise pertaining to emergency provisions, Disaster Imagination Game (DIG), and town watching to learn the safe and dangerous areas around their schools and school routes.

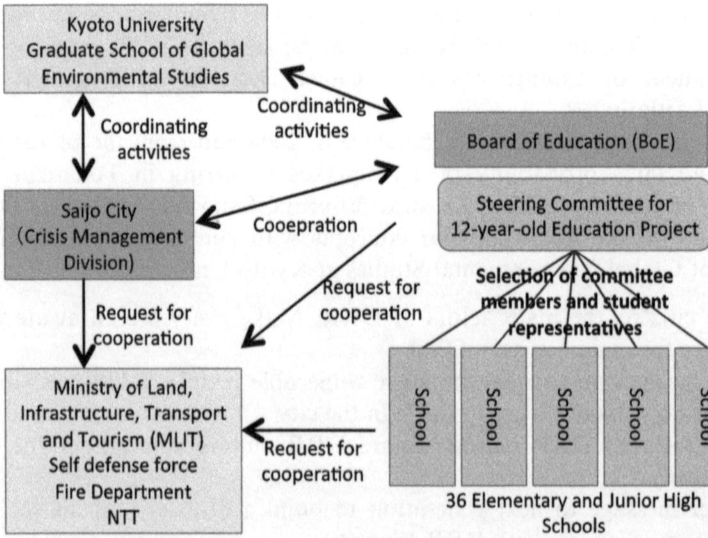

Fig. 1. Implementation Structure of 12-Year-Old Education Project.

The leaders who received the training then decide which activities will be implementable at school. Some of the activities conducted so far are making suggestions for better evacuation from the dangers of earthquake and tsunami, town watching of routes to school, and proposing emergency supplies that can be bought at 100 Yen store. Even adults are able to refer to the many ideas that were recommended from the DRR activities by the students.

After the EJET, the BoE interviewed 35 public schools and reexamined the evacuation location. Furthermore, the BoE inspected the time required to get to the secondary evacuation areas on foot and changed them as needed. Some of the schools, presuming emergency situations, added portable toilets and rearranged unused classrooms for evacuees when disasters occur. In order to ensure the sustainability of the 12-Year-Old Education Project, the "Forest is My Friend Project" is being implemented as follow-up training for junior high school students who have taken part in the 12-Year-Old Education Project. The city hopes that these students will be the central figures in taking actions during emergencies, even after they become adults. The DRR program does not aim to nurture DRR experts, but to raise consciousness that 6th-graders can also take important roles in DRR. Saijo has been attempting to build communities and networks through a very important festival, the Saijo Festival, which has existed since the Edo Period (1603–1868). Because the festival can be an opportunity to promote DRR, community DRR activity is conducted – the DRR Festival. Similarly, there is the Shishimai Festival, in which Shishimai from different regions gather to perform the traditional lion dance. Through these festivals, community-building and strengthening of ties can be achieved during normal times, leading to disaster resilience for the whole region. Saijo plans to continue with this kind of community-centered DRR awareness raising.

Experiences of Owase

Owase is historically a disaster-prone city located in the southern Mie Prefecture, along the Kumanonada seacoast. The city has experiences being affected by the 1944 Tonankai Earthquake and the localized torrential downpour in southern Mie in 1971. Aside from being in a region with the highest annual rainfall, and thus prone to water-related disasters, the city is also included in the designated area expected to receive extreme damage from tsunamis triggered by the anticipated Nankai Trough Earthquake. After realizing that the geographical features of the cities affected by the EJET very much resembled their own, Owase decided to wholly revise their DRR plans, including DRR education.

Prior to the EJET, the public schools in Owase have been conducting DRR education such as listening to experiences of past disasters, including the 1944 Tonankai Earthquake, watching videos of the Great Hanshin Earthquake and the 2004 Indian Ocean Earthquake and Tsunami, and incorporating the tsunami awareness story, "Inamura no hi" into DRR education. The evacuation drills were based on assumptions from past disasters and conducted with such instructions as "don't push," "don't run," and "don't speak."

Previously, the prefectural BoE conducted DRR education only for the designated schools, by dispatching earthquake simulation vehicles and providing lectures on liquefaction after earthquakes. Now, the BoE instructs all public schools to conduct DRR education. Owase has revised its tsunami evacuation for all elementary and junior high schools with the three principles below:

- "Don't believe in assumptions": Do not blindly believe the tsunami flooding areas marked on the hazard maps.
- "Try your best (to evacuate)": After evacuating to the designated area, continue to evacuate to a safer area if the situation allows.
- "Take lead in evacuation": Prepare to take the lead during evacuations.

The three principles above aim to build a safe city by developing a DRR culture that "Tsunami will come for sure, if you live in Owase" and by fostering the love of citizens for their hometown by stating, "Tsunamis will come to Owase occasionally, but Owase is a fascinating city." Each school established a "Working Group for Tsunami DRR Education Supervisors" that has been developing unified curricula for both elementary and junior high schools, regardless of the types of the anticipated tsunami and the size of the risk. The Working Group has also developed a DRR program that includes evacuation drills and designating new evacuation places and routes. Information on these activities are shared with the parents and local communities through the distribution of pamphlets (Fig. 1). Concerning the handover of students during emergencies, one of the key lessons learned from the EJET, the Working Group has been attempting to convince the parents that the schools will not hand over the students until safety from tsunami has been confirmed by the local authorities.

After the EJET, Yanohama Elementary School started to take measures to review and strengthen its DRR activities. As a first step, the school has changed its evacuation area from its sports field (9.1 m above sea level) to a nearby temple that is 23.5 m above sea level. In addition, awareness-raising and capacity-building were conducted for teachers to be able to make quick decisions according to the disaster situation to protect the safety of students, such as through DIG. Simultaneously, starting in 2012, DRR activities were marked clearly in the annual school activity plan and a teacher was appointed to be responsible for DRR education. Trial DRR classes using a "DRR Notebook" distributed by the prefectural BoE and a "Handbook for Tsunami DRR Education of Owase City" has been conducted in all classes. Joint evacuation drills are currently being planned so that the whole community will be able to take part in educating students to initiate and be responsible for their own safety. Remaining issues, such as roles in managing schools as evacuation centers and responsibilities for making decisions about student handover are still up for further discussions.

Experiences of Oobu

Oobu is a city located in Chita Peninsula Aichi Prefecture that possesses hillock and plain land no higher than 50 m above sea level. The city's goal is to become a City of Good Health that fosters health in both mind and body and to provide green and safe environments during natural disasters for its 87,374 citizens. In 2006, the city joined the Alliance for Healthy Cities of WHO. The city's history with natural disasters dates back to the Isewan Typhoon in 1959 and the Tokai Torrential Rain in 2000, which brought heavy water-related disasters caused by breaches of dykes and poor drainage. The city has also been designated as a "Region for the Advancement of Countermeasures against Disasters of Tokai, Tonankai and Nankai Earthquakes" that has been predicted with 88% probability to occur within the next 30 years. With this background, Oobu has been implementing DRR education in elementary schools as well as other DRR measures for the whole city.

Because of the anticipated Nankai Trough Earthquake, the prefectural BoE has been placing great importance on strengthening DRR education at schools. For this, changes in the perceptions of teachers about DRR must be first achieved. In the last fiscal year, teachers appointed for DRR from every public school (elementary, junior high, high, and special needs schools) were gathered to receive lectures on DRR, including a program by Gunma University. Furthermore, in coordination with Nagoya University and the prefectural disaster management division, the BoE has been training students from 15 high schools (5 participants each) in attempts to foster DRR leaders through DRR capacity-building through acquiring knowledge on natural disasters and building volunteer spirit during disasters.

Moreover, from the year 2012, in order to strengthen coordination among prefecture- and municipality-level schools and communities, the BoE has conducted workshops not only with appointed DRR schoolteachers, but also with municipal DRR officers as observers to discuss regional DRR measures and promote practical DRR education and management in accordance with the situation of each region. In the coming years, the BoE plans to expand and build networks with local governments and volunteer DRR groups.

Oobu has been implementing both structural and nonstructural DRR measures from its experiences of the Tokai Torrential Rain. For its structural measures, the city has installed 56 simultaneous transmission systems in strategic locations in the city and earthquake-proofed the gymnasiums of elementary and high schools that are designated as evacuation centers. For nonstructural measures, the city applied for funding for the "Regional DRR Model School Project" administered by the FDMA of the Ministry of Internal Affairs and Communications and has chosen two elementary schools to become model schools for DRR education. The project aims to build safe communities by connecting students who take part in DRR education and activities at the project model schools with their families and communities.

DRR education including the following has been conducted: DRR studies (mechanism of typhoons, taking action during water- and wind-related disasters, etc.), DRR drills (fire and earthquake evacuation drills and handover drills), DRR lectures for parents and communities, and DRR experience training (evacuation center experience, first-aid training, including use of Automated External Defibrillators). Many of these activities are implemented with participation from community members. Students over junior high school age were given additional tasks to help community people in DRR activities, including evacuation drills. A steering committee, consisting of teachers, a volunteer DRR group, the fire department, and local government (community safety division) was established to discuss on the operation of DRR activities and was proven effective to reach mutual understanding of the project objectives.

Good coordination has enabled the efficient implementation of different DRR activities in the city, such as conducting joint evacuation drills. For the first year, schools had difficulties in getting understanding from the community because the teachers lacked knowledge about implementing DRR education effectively. However, in the second year, the impact of the activities was gradually acknowledged, allowing further enhancement of these undertakings. For the third year, development of DRR programs for the next phase as well as increasing the model school to nine schools are being considered.

Around 80% of Oobu residents appreciate living in the city and wish to continue residing there. There is an increase of over 1,000 residents annually. The City government of Oobu believes that in order to achieve building the City of Good Health, an effective DRR plan is equally as important as fulfilling medical care and the social welfare of the city. Because of this, the city plans to continue implementing DRR activities in all elementary and junior high schools in the city. Coordination among the prefectural government, schools, families, and communities promotes mutual understanding about the importance of DRR and also the next steps for building a safe, ensuring community building.

ENHANCING LOCAL RESPONSES

After the EJET, on October 11, 2011, the Ministry of Education Sport Culture Science and Technology (MEXT, 2011) of the Government of Japan announced the concept of a "School Centered Community Building" for the recovery of schools that were affected by the EJET. The concept promoted three main components:

- securing land in safe location,
- strengthening DRR functions and eco-friendly measures, and
- making schools multifunctional community hubs.

However, specific activities at the local levels have not been realized, mostly because of complexities in the coordination of related government departments and building the consensus of communities. Up to now, most of the discussions have been concentrated on the hardware aspect of school recovery, but there is also a need to discuss nonstructural issues, such as rebuilding school–community linkage and the roles of schools in community-building.

Although rehabilitation and recovery of the schools must be promptly realized, school-centered community-building can become a much longer process, which could take 5 to 10 years or more. Therefore, this concept can be based on two different phases, a short- to mid-term recovery phase for rebuilding the school facility and a long-term phase for community-building. Transition between these two phases should be continuous and streamlined. However, as it is difficult for communities in the affected Tohoku area to think so far into the future and for other regions preparing for future disasters, such as the Tokai, Tonankai, and Nankai earthquakes, the ideas of community-building and the needs might change over the course of time, making sustainability the largest task in realizing this concept. Furthermore, as many regions in Japan, including the Tohoku Region, are facing the low birth–aging issues in addition to population drainage from the nonurban to urban areas, revitalization and survival of schools and communities must be holistically considered when planning recovery and community-building.

In order to place schools at the center of community-building, ways to link schools with communities must be considered. Education that places students and communities at the center is the key to this. Education based on local contexts allows students and communities to better understand, connect, and love their hometowns. This will encourage them to come back to their hometowns even if they were to leave their communities in the future. For example, deepening understanding of local economy through on-the-job training through school programs is an effective way to persuade students to become the community's future workforce, who would support the local economy.

Another way is to utilize local resources is by allowing communities to participate in educational activities. Each community usually has adults who have rich life experiences and an elderly group, who, in most cases, want to contribute to the community; these are human resource that the schools can rely on. The elderly people can also feel revitalized in having a role in contributing to the community and interacting with students. For example, the Aichi Prefecture has a concept, "the whole town is the teacher," in which the whole community is the classroom, the community members are teachers, and the school is the venue for community interaction.

In order for a school to function as a public facility that becomes the center of community-building, the prerequisite must be that the school will be open to the community. An idea for this is to open the school library to the community and make it a place for all community members to learn. Another option is to open up classrooms that are not being used to the community and make them places for community integration or temporary evacuation place during emergencies. The city of Kamaishi, for example, is currently considering rebuilding new schools, which were totaled by the disaster, to become multifunctional facilities that will be combined with community centers and preschools. The city says that this by itself is a contribution to community recovery and community-building.

In many regions, community members, including ones who do not have children in schools, participate in the school sports and culture festivals. In addition, schools can organize events intended specifically for community participation, some of which may be organized outside of schools to further community outreach.

In recent years, the Internet has become an effective medium for calling for community participation. In the city of Saijo, the administrators of a school originally planned to be closed took action to actively disseminate information about the school and the community, which led to an increase of students. In the age of low birth–aging society, school management has entered an era of competition in which the schools must actively promote their school as well as their community to attract students from local and external regions. This kind of appeal from schools will not only benefit schools, but will also contribute to the whole process of community-building.

At the time of the EJET, 89.3% (30,513) of the public schools were designated as evacuation centers, but most of these schools were not properly equipped to function as evacuation centers (equipped with storage room: 35.2%, water supply facility: 28.7%, electric generators: 18.0%, telecommunications equipment: 30.2%; based on an August 2011 survey conducted by the National Institute for Educational Policy Research (NIER, 2011). For schools to function as a central public facility, it is important that the school's DRR function is further enhanced for the community to acknowledge the school as a place to seek safety and assurance. Especially for small towns, the school is a symbolic facility in which they must operate as a place for safety during emergencies. Joint school–community DRR activities may be conducted to enhance the community's consciousness of school being a place to evacuate from dangers. This is something that the schools cannot do independently and will require coordination and information sharing with related departments of the local government and community organizations.

To understand the challenges and potentials of the school-centered community recovery, we organized a workshop of city officials and academicians and nongovernment organizations in 2012 in Kyoto, Japan. A group discussion was facilitated, which proposed the following two goals to achieve for school-centered community building (IEDM, 2013):

- To build communities centered around attractive and unique schools that incorporate regional characteristics
- To build schools with sufficient DRR functions so that the community can feel assured and safe from disasters

A few key factors to realize these goals are as follows:

- *Strong vision of leaders:* Strong vision and leadership of top officials and community leaders are essential in maintaining continuity of community-building activities.
- *Comprehensive coordination:* Community-building cannot be realized only by schools and the BoE; it requires assistance from other government departments, community members, and experts. Therefore, an organizational system, such as a steering committee, to coordinate stakeholders to work on a common task is necessary.
- *Build and strengthen partnership:* As community-building is a multidisciplinary process, support from universities and NPOs should be effectively utilized. For this, schools should actively seek partnerships with these organizations.

Sustainability is the key for a successful school-centered community building. For the sustainability of a school-centered community building, the following actions need to be taken:

- *Child- and community-centered education and school management:* During the recovery and community-building process, reassignment of schoolteachers and BoE officials is inevitable and because of this, they are usually preoccupied with work that they can do within their tenures. Therefore, in order to maintain the continuity of the process, it is essential to develop a system for students who will stay in the communities to take a larger role in educational activities and school management.
- *Maintain and reinforce existing system:* Another key to sustainability of community-building is to maintain and reinforce existing systems together with new efforts.
- *Developing an evaluation system for people working in recovery and community-building:* Developing an evaluation system can become a strong incentive for people involved in recovery and community-building. Establishing a certification system for DRR education and regional education could also help with this evaluation.
- *Develop a system to evaluate the stages of recovery and community-building:* Evaluating and publicizing the process of recovery and community-building and marking the milestones can help in drawing up detailed implementation plans. Particularly in the disaster-affected areas, the changes in the recovery process and needs are dynamic, hence making long-term planning difficult. This adds to the worries of the people who were affected by the disaster.
- *Exploring possibilities of community business:* One effective method that could ensure the sustainability of community-building activities that can be considered is establishment of community businesses. In the initial few years, recovery budgets or external funding may be used, but eventually, community-based NPOs can carry on with the community-building activities as business.
- *Being critical about school and community survival:* When a school scales down or closes, communities will lose their children. Inevitably, decreasing numbers of children will advance aging in the communities. In Saijo, the mayor has declared that the schools in the city will not be closed under any circumstances because when schools close, the community might also vanish. By sharing this consciousness with the community, the city hopes to maintain its efforts to actively link schools and communities.

ACKNOWLEDGMENT

We acknowledge support from CWS [Church World Services] Asia Pacific and MERCY Malaysia for conducting this study.

REFERENCES

Arya, S. A. (1993). *Training and drills for the general public in emergency response to a major earthquake*. Training and Education for Improving Earthquake Disaster Management in Development Counties, UNCRD Meeting Report Series, No. 57 (pp. 103–114).

Cabinet office (CAO). (2011). *Government of Japan, Central Disaster Management Council.* Retrieved from http://www.bousai.go.jp/jishin/chubou/ toshibu_jishin/3/4-1.pdf. Accessed on November 10, 2011.

Frew, L. S. (2002). *Public awareness and social marketing*. ADPC regional workshop on Best Practices in Disaster Management, Bangkok (pp. 381–393).

IEDM. (2013). *Workshop on enhancing disaster resilience of education sector and communities.* Kyoto University International Environment and Disaster Management (IEDM) Publication, Kyoto, Japan (52pp.).

Kuroiwa, J. A. (1993). *Peru's national education program for disaster prevention and mitigation (PNEPDPM)*. Training and Education for Improving Earthquake Disaster Management in Developing Countries, UNCRD Meeting Report Series, No. 57 (pp. 95–102).

Ministry of Education, Culture, Sports, Science and Technology (MEXT). (2011). *Urgent recommendation "Concerning school facility improvement in light of the damage caused by the Great East Japan Earthquake"*. Government Report, Tokyo, Japan.

National Institute for Education Policy Research (NIER). (2008). *Gakkou shisetsu no koujyou no tameni.* (In Japanese). Retrieved from http://www.nier.go.jp/ shisetsu/pdf/bousaitsuiki.pdf. Accessed on November 15, 2011.

National Institute for Education Policy Research (NIER). (2011). *Gakkou shisetsu no bousaikinou ni kansuru jittai cyousa ni tsuite*. (In Japanese). Retrieved from http://www.nier.go.jp/shisetsu/pdf/bousaikinou2011.pdf. Accessed on March 15, 2013.

Okada, T. (2008). *Anzen anshin de yutakana gakkou shisetsu dukuri* (In Japanese). Kyouiku Inkai Geppou, No. 700 (pp. 7–10).

Radu, C. (1993). *Necessity of training and education in earthquake-prone country*. Training and Education for Improving Earthquake Disaster Management in Developing Countries, UNCRD Meeting Report Series, No. 57 (pp. 15–33).

Shaw, R., & Kobayashi, M. (2001). The role of school in creating earthquake-safe environment. Paper presented at the OECD Workshop Disaster Management and Education Facilities, Thessaloniki, Greece (pp. 7–9).

Shaw, R., Shiwaku, K., Kobayashi, H., & Kobayashi, M. (2004). Linking experience, education, perception and earthquake preparedness. *Disaster Prevention and Management, 13*(1), 39–49.

Shaw, R., Shiwaku, K., & Takeuchi, Y. (2011). *Disaster education* (p. 162). Bingley, UK: Emerald Publisher.

Shiwaku, K., Fujieda, A., Takeuchi, Y., & Shaw, R. (2010). Utilization of disaster experiences in school disaster education in disaster affected area. *Japan Society for Natural Disaster Science Journal*, *29*(1), 83–95.

Suda, Y. (2012). *Linking human behaviour, participatory mapping and community infrastructures for disaster risk reduction: Approaches in Shiso and Kamaishi*. Master thesis, Kyoto University, Kyoto, Japan.

CHAPTER 10

RISK COMMUNICATION THROUGH COMMUNITY-BASED SOCIETY ORGANIZATIONS AS LOCAL RESPONSE TO DISASTER IN BANDUNG, INDONESIA

Farah Mulyasari and Rajib Shaw

ABSTRACT

This chapter illustrates local responses to disaster and highlights the potential role of three community-based society organizations (CBSOs) – women's groups, youth groups, and religious groups – as risk communicators in Bandung, Indonesia. A framework is modeled for CBSOs' risk communication process in bridging the gap between the local government and the community. A set of indicators in social, economic, and institutional resilience activities (SIERA), with a scope of 45 disaster risk reduction (DRR) activities covering three different disaster periods was developed to characterize the process of the delivery of risk information by these CBSOs through their activities at subdistrict and ward levels. The data were collected through a questionnaire survey method using the SIERA approach. Each CBSO leader in a ward was

Risks and Conflicts: Local Responses to Natural Disasters
Community, Environment and Disaster Risk Management, Volume 14, 231–250

ISSN: 2040-7262/doi:10.1108/S2040-7262(2013)0000014016

surveyed about their perceptions of these 45 ongoing SIERA activities and their risk information source and dissemination process. Statistical analysis was applied to determine the relationship between variables such as periods of disaster and types of SIERA activities and its attributing factors (location, population, and dynamic of organizations in their locale) in finding variations of risk communication activity that may function for communities. Five risk communication processes of the CBSOs are identified; when their perceptions and ongoing activities are compared, activities such as dissemination of disaster risk information, conveying early warnings to their peers, and involvement of the local government have been carried out by these CBSOs. This indicates that CBSOs' activities already have a certain degree of risk communication embedded in the communities. The results confirm that these CBSOs, through their social networks, can become active agents of change and bridge the communication gap between government and community. Thus, CBSOs' risk communication provides the opportunity to contribute to the overall resilience-building and disaster risk reduction as part of people-centered actions and local responses to disasters.

Keywords: Community-based society organizations; people-centered actions; religious group; risk communication; women group; youth group

INTRODUCTION

Throughout history, human beings have been self-reliant and supported by near relatives only in coping with adverse situations, including disasters (Helsloot & Ruitenberg, 2004). This type of coping with disaster is manifested in several responses, depending on the disaster time period (before, during, or after) as well as on the disaster type (natural (geological and climate-related, including hydro-meteorological), man-made, or technological). The natural type of disaster will be discussed in this chapter. Leitmann (2007) gave an example of post-disaster response in which cities in Indonesia responded in terms of public health to recent urban-centered calamities. Indonesia has been struck by a series of major disasters – the 2004 tsunami and earthquakes (devastating the cities of Banda Aceh, and Meulaboh) and the 2006 central Java earthquake (impacting the city of Yogyakarta and its suburbs). Moreover, floods have been widely and frequently occurring in Indonesia. The 2002, 2007, and 2012 Jakarta floods had a massive impact on the city and its citizens, causing disruption in socioeconomic and institutional terms (IFRC, 2007; Mulyasari, Shaw, & Takeuchi, 2011; Steinberg, 2007). The aforementioned disasters call upon responses and countermeasures, which have to be beneficial to communities, as they are the first and most impacted by disasters.

In order for proper responses to a disaster to be taken, communities need information about how they are at risk, and they need to know who is the sender of this information; thus, risk communication is crucial at this stage. By providing the community with information, risk communication becomes two-way communication, which involves providing information, understanding people's perception of the risks, and developing solutions in partnership (Mulyasari et al., 2011). Aside from the government, community-based society organizations (CBSOs) as part of civil society organizations, such as women's groups, youth groups, and faith-based organizations, can be one of the main actors in communicating the disaster risk to the communities. They possess baseline characteristics that highlight the participation of community. They are effective in bringing about community change, including disaster risk reduction (DRR) within the community at large, through its collective action on shared interests (Mulyasari & Shaw, 2012). They no longer provide services that government has failed to reach, but they are now more in the mainstream of development activities (Clayton, Oakley, & Taylor, 2000). Being close to the grassroots, their scale and profile of community-based DRR (CBDRR) have been acknowledged and recognized by the government. This recognition enables them to mainstream and conduct activities in reducing risks that target the beneficiaries for communities (Mulyasari & Shaw, 2012). Thus, the risk communication process implemented by these CBSOs will impact the communities positively and ensure that response actions are participatory. An additional value that these CBSOs can bring as the risk communicators is that they could bridge the communication gaps between government and the communities, as the main source of risk information.

This chapter aims to illustrate the example of local responses to disaster and highlights the potential role of the three CBSOs as risk communicators in Bandung, Indonesia. A framework is modeled for the CBSOs' risk communication process in bridging the gap between the local government and the community. A set of indicators in social, economic, and institutional resilience activities (SIERA), with a scope of 45 DRR activities covering three different disaster periods, was developed to characterize the process of the delivery of risk information by these CBSOs through their activities at subdistrict and ward levels. The data were collected through a questionnaire survey method using the SIERA approach. Each CBSO leader in a ward was surveyed about their perceptions of these 45 ongoing SIERA activities and their risk information source and dissemination process. Statistical analysis was applied to determine the relationship between variables such as periods of disaster and types of SIERA activities and its attributing factors (location, population, and dynamics of organizations in their locale) in finding variations of risk communication activity that may function for communities as responses to disaster at the local level.

CBSOS AS LOCAL RESPONDERS TO DISASTER IN INDONESIA

Local responders to disasters in Indonesia are fundamentally the communities, represented by their three biggest organizations – women's groups, youth groups, and religious groups (mosque leaders). These are the CBSOs, the name of which is derived from the term "civil society." The definition of *civil society* has been given in many references. According to the Kaldor (2003, p. 11), "civil society could be described as those organizations, groups and movements who are engaged in this process of negotiation and debate about the character of the rules – it is the process of expressing 'voice.' " The civil society commonly embraces a diversity of spaces, actors, and institutional forms; varying in their degree of formality, autonomy, and power. Civil societies are often populated by organizations such as registered charities, developmental nongovernmental organizations (NGOs), professional associations, and community groups, such as women's, youth, and faith-based or religious organization. Those groups are seen as the nearest to the grassroots level and therefore could best accommodate the aspirations and needs of the communities. Kamat (2003) states that CBSOs are locally based organizations seen as the champions of "bottom up" or "pro-people" development. Moreover, CBSOs have an active membership base among the particular community to which they belong, be it urban or rural. These "target" or "client" groups at the local level are themselves involved in decision-making processes and provide organizational direction that engages all other community members.

As stated above, CBSOs possess baseline characteristics that highlight the participation of different types of community groups. Those groups are considered to be effective in bringing change within the community at large, since its nature is that of collective action around shared interest.

Bringing change within the community includes responding to disasters locally, such as through risk reduction and communication. Combining the nature of CBSOs and DRRs, then CBDRR is shaped to fulfill the needs of and deliver services and mainstream DRR activities to communities. On the one hand, providing social services has been a critical role traditionally played by CBSOs (Clayton et al., 2000), but on the other hand, the key change that has taken place in recent years is that CBSOs are no longer just providing services to people that the government has failed to reach, but they are now far more in the mainstream of development activities. Below are the illustrations of each aforementioned CBSO.

Women's Groups

The women's groups that is rooted in the community is the Women Welfare Association (WWA) – more well-known as TP PKK (Tim Penggerak Pemberdayaan Kesejahteraan Keluarga). This is classified as one of the major CBSOs in Indonesia and has chapters spread throughout the country. These chapters are established in different administrative levels of governance hierarchy, from national to the smallest administrative units, such as wards (in cities) and villages (in rural areas). At the national level, the WWA is headed by the wife of the Minister of Home Affairs, at the provincial level by the wife of the Governor, at the city level by the wife of the mayor, at the subdistrict level by the wife of the head of the subdistrict, and at the ward level by the wife of the ward leader. These positions will change once every five years, in line with their husbands' regional leadership elections.

The WWA is unique; the women work autonomously as an independent and nonprofit social institution not affiliated with any particular political party; yet it is subsidized by the government. Referring to their historical roots in 1967, according to the Central Secretariat of TP PKK (2006), the WWA was established to improve the condition of Indonesian families. More than 50% of Indonesia's population is women, and especially those who are living in rural areas have low economic status and education levels, causing high rates of maternal and infant mortality. Recognizing this, the WWA moved to increase its activities, particularly for the efforts of decreasing the high maternal and infant mortality rates. Seeing the unfortunate condition of the people of Central Java Province back then, the wife of the Central Java Province Governor, pioneered in the establishment of the WWA as the women's movement that aims to raise the welfare and awareness of families. In order to achieve its vision, the WWA has adopted 10 main programs, including the efforts to meet basic human needs – fulfilling the physical, mental and social aspects. In its most recent developments, the WWA is also engaging in environment-related issues, such as sanitation and disaster relief. Consequently, the WWA is highly regarded as a potential driver and change of agent for DRR activities within the communities. It is seen not only from their possible programs, but also from the women's spirit and strong will to improve their lives, families, and surrounding community (Mulyasari, Shaw, & Takeuchi, 2015).

Youth Union

Youth Union is a youth movement that is actively engaged in the communities. It is established by the Ministry of Social Welfare of Republic Indonesia throughout Indonesian cities and villages; the ages of its members range from 13 to 45 years (Ministry of Social Affairs, 2010). The Youth Union (YU), or the Karang Taruna, is coupled with programs of and working in the field of social welfare and community development. As a community organization, the YU has a board of management and voluntary members. The members carry out their functions in their respective sectors and are able to work together with the local administration. Their activities' programs are in accordance with the needs, problems, and issues revolving around youth and communities.

In the urban areas, one of the YU's strengths is the ability to raise funds that come from the government, NGOs, the private sector, and from the community itself. As an example, for events such as yearly national independence commemoration day (August 17) and religious festivities, the youths within the ward gather and develop plans in organizing and holding the event. It starts by raising funds from the community, private companies, NGOs, and so on, and then organizes and implements the event for the community. The youth within the YU are considered to be the engine propelling the activities within the community. In the rural areas, the youth apply a different approach to serve and participate in the rural community development, especially in the field of raising the social welfare. One of many important roles that the youth play in rural development is the community empowerment organization (LPM/*Lembaga Pemberdayaan Masyarakat*). The LPM is a public institution, representing the aspects of all communities in the rural development and having the task of organizing meetings in the village; hence, the YU coordinates, consults, assigns tasks, and gives suggestions together with the LPM. Thus, the YU's main task is empowering the youth to address social welfare issues with the local government, as a form of community participation (Ministry of Social Affairs, 2011a).

The YU participates in the community by volunteering, in the form of delivering social services and rehabilitation, such as environmental protection and sanitation and supporting vulnerable groups such as children, disabled, women, elderly, and disaster victims. At the national level, YUs explore and perform other socioeconomic activities, such as economic development through cooperation with existing social organizations (e.g., the formation of the joint business group), enhancing skills, and entrepreneurship. The Ministry of Social Affairs has formed TAGANA (Taruna Siaga Bencana/Youth Disaster Response), with more than 30,000 people throughout some of Indonesian provinces, which is presented with specified targets for disaster emergency and response. TAGANA was established by the Ministry of Social Affairs in 2009 as government's response after the 2004 Indian Ocean Earthquake and Tsunami and the 2006 Yogyakarta Earthquake. The members of TAGANA are mostly the youth in YUs and is not limited to any specific age range (Ministry of Social Affairs, 2011b). TAGANA is actively engaged with the communities and local authorities in responding to situations during and after a disaster. Regardless of the type of disaster, activities such as search and rescue, setting up community kitchens, collection of data with regard to disaster losses, and informing local authorities (wards and subdistricts) and local disaster management agencies are considered as disaster responses of TAGANA. TAGANA is equipped and mobilized only if a disaster occurs.

Faith-based or Religious Organizations

Many religious organizations or groups exist in Indonesia; some of them have a strong and long historical background, dating from colonial times. Many of them were established as a means of promoting social welfare as well as education of their members. Several of them have strong roots among the population that spread to most parts of the country. In many cases, they have charismatic, respected leaders who act as informal leaders within the communities and often act as the counterpart of government officials in daily social and political activities. They are often considered to have a strong influence in implementing responses to disaster, especially disaster reduction awareness programs, among their followers.

Implementing CBDRR activities among communities can be considered an opportunity, as most religious organizations have been involved in many disaster relief and recovery activities during disaster situations. Partnering with an academic or research institution as their technical support can provide knowledge and skills to develop potential capabilities in conducting and implementing CBDRR programs.

Nahdlatul Ulama and Muhammadiyah are the two largest Islamic organizations in terms of members and spatial distribution. They have been implementing CBDRR activities involving schoolchildren and pesantren (Islamic boarding school) students, youth Islamic organizations, and communities around pesantren within mosques. A recent study by IRI, UNOCHA, & LPBI NU (2011) about the potential role of the mosque in disaster situations in Indonesia, shows that mosques are coming to perform an important function for surrounding communities in providing services during disasters. Nahdlatul Ulama mobilized Islamic boarding schools (more than several thousand spread out around Indonesia) as their capacity-building target, because these boarding schools are considered as the local center in providing and conveying disaster risk information to communities. Muhammadiyah has worked closely with elementary schools and youth organizations under its jurisdiction for with regard to disaster awareness programs. According to CDASC (2008), children and youth represent the future generation and act as ambassadors or media to communicate the disaster risk information to their parents and other family members. Therefore, these entities (Islamic boarding school students, children, and youths) act as agents of change and development in building the safety culture among their communities.

Consequently, these two religious organizations are making an effort to build disaster risk awareness among the community as well as among their members. Thus, organizing community workshops and training and conducting participatory activities related to understanding hazards in the area, demand action in addressing the pressure of increased numbers of disaster events in Indonesia. The groups facilitate bridging the hazard knowledge gap of the people and develop common sense procedures for safety both in rural and urban areas. Both areas are susceptible to various hazards because of a lack of knowledge, lack of access to infrastructure and services, and lack of opportunity and resources, which impede the implementation of DRR initiatives.

Another type of faith-based or religious organization is DKM (Dewan Keluarga Masjid/Mosque Council), which can be found in any major mosque in every city, subdistrict, ward, and village. These Mosque Councils are independent, yet their organizational structure is always headed by men and mostly by the elderly or a community leader. Their activities are mainly fund-raising for the poor and organizing religious events, such as providing the religious materials and educating youngsters as well as holding a men's and women's forums for conveying and discussing not only religious matters, but also for addressing social issues. Three types of CBSOs are shown in Fig. 1(a–c).

CASE STUDY: BANDUNG, INDONESIA

The potential of CBSOs in responding to disasters are in the form of CBDRR. These DRR activities are reflected in day-to-day activities, related to their regular programs and the environment that impacts communities to a large extent. To identify what kind of local responses to disasters are implemented by these CBSOs, an initial capacity assessment was performed to find out the resilience level and to delineate areas of improvements of the city at the subdistrict level.

Bandung City Government carried out the assessment in the form of climate and disaster resilience mapping. The city utilized the Climate-Disaster Resilience Index (CDRI) method developed by Joerin and Shaw (2011). The CDRI is a tool to measure the resilience of a system (city with its government, institutions, infrastructures, and communities) with regard to climate-related disasters. The CDRI provides a comprehensive baseline assessment that addresses the linkages of the system – those between

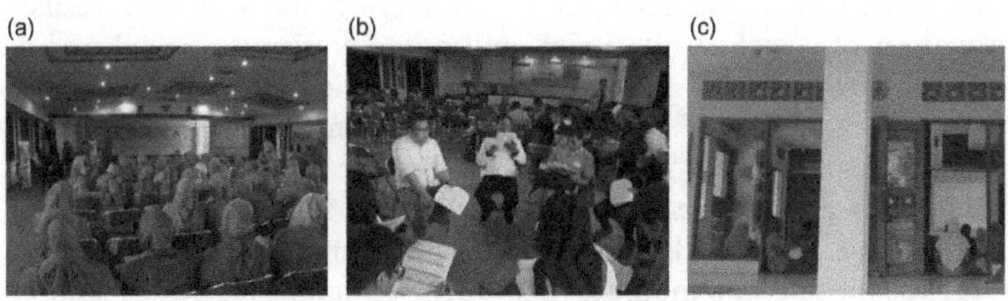

Fig. 1. (a) Heads of Women Welfare Associations of Wards of Bandung Attending Sanitation and Women and Climate-Related Disaster Workshop; (b) Members of Youth Unions of Bandung Discussing Youth Risk Communication; (c) Members of Mosque Councils of a Ward in Bandung Attending a Community Meeting in a Mosque. *Source*: Authors.

various aspects of actors and components of the city. This resilience assessment looks at 5 dimensions and 25 parameters. Resilience is important in the DRR approach because it measures the capacity of a system – such as a city system that has various forms of dimensions (physical, social, economic, institutional, and natural resilience) – to respond and recover from shocks and stresses. Once the resilience is measured, targeted and specific response actions could be then implemented to enhance resilience and at the same time reduce risks.

Based on the assessment, the overall CDRI values (1 = poor to 5 = best) of the subdistricts are derived. The lowest resilience values are scored by the poor areas as well as by the socially disadvantaged; specifically, economic resilience is indicated by budgets and subsidies for disaster risk management and reduction and natural resilience by the frequency of natural hazards. The highest resilience values are marked for social resilience in the health sector; for physical resilience by electricity, water, and accessibility of roads; and for institutional resilience by institutional collaboration with other organizations and stakeholders during a disaster. The rest of the majority of subdistricts have similar middle-class resilience values, for which the variations in the scores are very small. All the resilience values have similar values to each other and provide a potential opportunity to improve from the same starting point to take off and grow collectively toward the enhancement of Bandung climate-related disaster resilience (Shaw, Takeuchi, & Mulyasari, 2012).

The last part of the resilience assessment is the communication – how the risk and resilience information collected at the root level are conveyed to wider communities, which would result in their responding properly to a disaster. For that, the social, institutional, and economic resilience actions (SIERA) is the medium for conveying the risk and resilience information of social, institutional, and economic issues. Social and economic resilience are the most important determining factors of resilience that lead to a course of sustainable development. Social resilience is an important component in circumstances under which individuals and social groups adapt to environmental change. Adger (2000) opines that ecological and social resilience may be linked through the dependence on the ecosystems of communities and their economic activities and that these make social and economic resilience one whole undeniable package. Moreover, social resilience is institutionally determined, in the sense that institutions permeate all social systems and institutions thereby fundamentally establish the economic system in terms of its structure and distribution of assets.

The SIERA approach integrates these three dimensions. It analyzes the DRR activities through primary indicators for the phases before, during, and after a disaster to describe different types of community groups' coping strategies during the whole disaster cycle. McEntire (2001) emphasizes that a holistic approach to the disaster problem is needed, and approach in which disaster agents no longer are limited to natural and technological hazards. In addition, disaster management should no longer imply only a response attitude. What is needed is an approach that addresses all agents, all actors, and all phases pertaining to disaster vulnerability. In this respect, the scope sets the course of plausible CBDRR activities to be implemented by the CBSOs at the city, subdistrict, and ward levels. Each scope of CBDRR of each CBSO corresponds to 3 dimensions (social, institutional, and economic) and 15 primary indicators of the CDRI (Table 1).

The CBSOs' resilience activities (hereafter called CBSOs' SIERA) were developed through literature reviews about what kind of DRR activities may lead the community to become resilient to climate-related disasters. The activities of three CBSOs in particular were obtained from field observations and informal interviews with their members – Women Welfare Associations, Youth Unions, and Faith-Based Organizations in Bandung. Moreover, local experts' view were acquired from women's, youths', and religious' leaders at various levels in the city.

ADVANCING CBSOS' SIERA OF BANDUNG AS A RISK COMMUNICATION PLATFORM

CBSOs' SIERA are embedded in the communities' daily lives and working programs. It provides the opportunity and entry point for the risk communication process as local responses to disasters. These actions are done hand in hand with communities in their neighborhoods. When implementing these, both parties, whether the CBSOs of women's, youth's, or faith-based groups, together with the community, are shaping two-way risk communication. According to the National Research Council (1989), the risk communication process is an interactive process of exchange of information and opinions among the individuals, groups, and institutions. It is a dialogue in which multiple messages are discussed. Thus, when these CBSOs are performing their SIERA, not only are they exchanging information about disaster risks, but they are also exploring collaborative solutions.

Table 1. CBSOs' (SIERA) Derived from CDRI.

	CBSOs' DRR (resilience) activities in *SOCIAL* issues (*before, during, and after a disaster*)
Scope	*Population, health, education and awareness, social capital, community preparedness during a disaster*
	Example: Women's groups: conduct health campaigns, pass on emergency warnings, provide education and access to resources, etc.
	Youth groups: disseminate disaster awareness information and conduct drills, evacuate vulnerable groups to safer areas, clean up the debris after a disaster
	Faith-based groups: list most vulnerable households, collect used clothes and goods for donation, organize community in distribution of aids
	CBSOs' DRR (resilience) activities in *INSTITUTIONAL* issues (*before, during, and after a disaster*)
Scope	*Mainstreaming of DRR and CCA, Effectiveness of crisis management, knowledge dissemination and management, institutional collaboration, good governance*
	Example: Women's groups: train volunteers, network, gather disaster information, etc.
	Youth groups: mobilize youth in volunteering training and first aid, gather information about disaster loss for authorities
	Faith-based groups: disseminate printed safety materials after praying sessions, mobilize men's community members in neighborhood watches, engage community leaders in the development of rehabilitation and reconstruction process plan
	CBSOs' DRR (resilience) activities in *ECONOMIC* issues (*before, during, and after a disaster*)
Scope	*Income, employment, household assets, finance and savings, budget an subsidy*
	Example: Women groups: provide small-scale insurance, inventory destroyed and intact household assets, link local government in rehabilitation of education sector, etc.
	Youth groups: engage private sector in giving youth entrepreneurship, fund-raise, engage outside experts in providing youth skills in setting-up home industries
	Faith-based groups: link households to religious-based bank institution for saving practice, fund-raise during/after praying sessions and distribute funds, link local government in rehabilitation of mosques, community halls, and schools

For example, the women's groups in Bandung have different ways of responding to a disaster (mostly flood/inundation). Their role centers mainly around organizing the community kitchen and organizing training for the community, such as flood preparedness and awareness. For example, after the heavy rainfalls in 2005 and April 2013, their area was badly inundated, the women's group in subdistrict XV of Bandung regularly invited experts and organizations such as firefighters and local government agencies to give women and their communities knowledge about flood/fire mitigation. In addition, women's group informed the community about making bio pore absorbing holes to let the rainwater be absorbed by soil. Bio pore is an alternative to the absorbing well for absorbing rainwater; it is called "the palace for worms" because of its organic content (rubbish) so it will lure insects (ants, worms, termites, etc.) into the hole and making bio pores as small tunnels so that the water can be absorbed sooner. The making of bio pores during the rainy season has been endorsed by the head of the women's group of Bandung as their routine program in response to floods. Another type of women's group responds to disaster by mobilizing and encouraging the community in volunteering to undertake actions when a disaster occurs (Mulyasari et al., 2015). Moreover, based on SIERA survey results, these women's groups are mostly active before a disaster. Their actions in health issues address programs in line with Bandung government's policies and priorities. More than 70% of the wards conduct monthly health campaigns that are organized by these women's groups. The campaign targets women's health, the development, growth, and nutrition checkups of babies and small children, and health and nutrition control among the elderly. Thus, aside from community-preparedness actions, addressing health issues signifies women's groups' responses to disasters.

The youth groups in Bandung, in the form of unions in wards and neighborhoods, have mobilized themselves to be engaged in disaster issues. Aside from youth unions' activities, such as fund-raising and mobilizing the men's community members to be part of community security watches, the TAGANA as Youth Disaster Response group is involved mainly during disaster situations, such as emergency responses. For example, in 1983, Bandung experienced a flood and landslides in its northern part because of heavy rainfall; many households were inundated, and the disaster caused human casualties. Not only the youth union in this region responded to this disaster, but also youth unions from other regions responded. Their responses were in the form of voluntarily giving direct support in SAR (Search and Rescue) activity and first aid. For this purpose, the youth union had informed and engaged TAGANA and the Red Cross to support the youth union in responding to this disaster. The responses of youth unions are mostly in physical activities. Moreover, the outcome of SIERA survey results show that when their perceptions and ongoing activities are correlated, a very high degree of relationship/correlation ($r^2 = 0.94$) is disclosed. Since youth unions and TAGANA of Bandung have a close working network with local government and agencies, their SIERA is dominated by institutional issues. It is also worth mentioning that youth unions in Bandung established the Productive Economic Business program, which is based on private companies' donation to improve youth's economic condition in neighborhoods and bring alternative income to households. Making stickers and t-shirts and selling prepaid mobile phone vouchers are their main business (Head of Youth Union of Bandung City, personal communication, November 2011). Another finding from YUs' SIERA approach is that youth unions located in different parts of the city (such as the center, the residential areas, and the outskirts) respond to disaster differently. Therefore, their risk reduction and risk communication actions vary from one area to another.

The mosque councils as faith-based organizations in neighborhoods have responded differently to disasters than have women's and youth groups. For example, during the flood in 2005, the mosque councils in this region informed and coordinated with the local market in their neighborhoods to monitor water sewage. They mobilized the surrounding community in doing the "gotong royong" (Indonesian custom/spirit to work together) in the removal of waste from the sewers and ditches. The mosque councils coordinated the fund-raising and charity works to collect support (food, donation, used goods) to be distributed to the affected community. Moreover, through the SIERA approach, the mosque councils took part in risk reduction and risk communication actions. However, among these activities, strong values for action in the social dimension are coming up from these organizations at 150 wards in Bandung. These show that faith-based organizations have provided communities in neighborhoods a place to gather and interact. The regular meeting for prayer (five times/day) is the entry point where communities can meet, talk, and discuss issues such as disaster. Therefore, faith-based organizations provide the community a risk communication platform. In contrast, other results from SIERA indicate that mosques would like to contribute to more-viable economic activities in the long run. This highlights that faith-based organizations could provide economic perspective to the community. Therefore, being a center of community activities in the neighborhood, faith-based organizations can enhance the economic resilience of the community. In terms of responses, faith-based organizations in Bandung have a high degree of correlation between actions during and after a disaster. Faith-based organizations need to be more engaged in risk reduction and risk communication actions before a disaster, such as disaster education and awareness. All in all, these activities could be drawn as their responses to disasters. A summary of the responses by these three CBSOs are illustrated in Table 2.

POTENTIALS AND CHALLENGES OF CBSOS' RISK COMMUNICATION AS LOCAL RESPONSE TO DISASTERS

There is no doubt that CBSOs are important in contributing the DRR actions as local responses to disasters. Victoria (2009) emphasizes that whether a disaster is major or minor, of national or local proportions, it is the people in the communities who will suffer most from its adverse effects. People, collectively, use coping and survival strategies to face and respond to the situation long before outside help from the government or NGOs arrives. There are challenges and potentials with regard to how to more effectively advance CBSOs as local responders to disasters. The involvement of community members in CBSOs is to provide mutual support and opportunities in delivering services such as risk information and lobbying the authorities to accommodate their needs and concerns. Therefore, community participation is one of community empowerment strategies (Kieffer, 1984; Paton & Bishop, 1996). This is a crucial element in risk reduction and risk communication actions.

The CBSOs encourage potentials as local responders because of their social capital and tightknit networks with the community. For example, the responses by the women's, youth, and religious group, such as informing, engaging, and mobilizing the community and other stakeholders to take part in DRR as well as in actions during emergencies, show that they can persuade and gain trust from the community. In addition, the spirit of "gotong royong" has added value to their social capital. Putnam (2000) discuss how social capital provides informal educators with a rationale for their activities. The classic working environment for the informal educator is the group, club, or organization. The interaction within the CBSOs enables communities to build upon them, commit to each other, and knit the social fabric together. Because of their close relationship with the communities, they obtain trust, which is an important element in risk

Table 2. Summary of CBSOs' Responses.

Characteristics	Types of Responses		
	Women's groups (Women Welfare Associations)	Youth groups (Youth Unions)	Faith-based groups (Mosque Councils)
Type of organization	Semi-formal – Headed by the wives of appointed administrative leaders.	Volunteer – Headed by appointed youth role models at city level and selected at lower-level administrative units (subdistricts, ward, neighborhoods).	Volunteer – Headed by a community elder/ community leader.
Timing of response	Mostly before a disaster	Mostly during a disaster	Mostly after a disaster
Nature of actions	• Actions planned in working period (short and long term); primarily active in social aspects, especially in health issues.	• Actions mostly ad hoc.	• Actions dominated by humanitarian aspects, such as collecting goods, used clothes, and foods for donation to the poor and refugees; fund-raising during prayer sessions and distributing collected funds to the needy.
	• Utilize strong networking among the women and community members. • Have the advantage of direct relationship with the local government (at city, subdistrict, and ward levels); risk communication actions as local responses to disasters are better rooted in the community.	• Because of TAGANA, youth unions have close networking with local disaster agencies and interdepartmental agencies. • Mobilize TAGANA (youth disaster response) units have influenced the youth in taking risk communication action in emergency responses.	• Utilize the medium of face-to-face interaction with the community in the prayer sessions (five times daily and once a week), allowing councils to have direct communication contact and frequent access to community to inquire about the needs and concerns after the disaster; respond collaboratively.

communication, but it may also be considered a challenge. This social trust is conceptualized and advocated by Cvetkovich and Earle (1995, in Cvetkovich, 2000). This approach rests upon the idea that people's trust in risk-managing institutions or organizations is a major mediator of their risk-related perceptions and behaviors. Trust, for example, is regarded as vital for public acceptance of risk management and its authority. In other words, communities are more likely to follow the advice of risk-managing bodies or organizations if they perceive them as trustworthy (Covello & Sandman, 2001). Whether communities place trust in authorities hinges on whether they deem institutional actions to be based on the same values as they hold themselves. It is on these terms that communities grant institutions or organizations the responsibility of managing risks on the communities' behalf. The main implication for risk communication is that it should aim to build trust in risk-managing bodies. Open and honest communication that follows the principle of transparency is widely held to be key in promoting trust that these organizations will respond properly to disasters. To keep the trust of the community is surely a challenge for the above three CBSOs in executing their responses to disasters. The earlier examples show that in order for the risk reduction/communication action to be successfully implemented, those CBSOs must already have obtained a certain level of trust from their neighborhoods. Thus, maintaining the community's trust is a prescription for CBDRR as a local response to disasters.

Lastly, another challenge is risk governance/management. Communication happens at different management or governance stages and is characterized by different aims and priorities. The need for a clear distribution of responsibility among actors is key in order to have one stream of information both among responsible bodies and toward the population, to avoid repetition and contradictory information, and to have an effective chain of delivery (Höppner, Brundl, & Buchecker, 2010). This is true for warnings during an emergency and for long-term communication that aims to build up a culture of risk within a community. The challenge for the three CBSOs, based on examples of their responses as discussed above, is that they should have clear guidelines and task descriptions with regard to who is doing what. The description need not be a formal one, but it should be informally agreed upon among their members. Particularly in emergency situations, the challenge with communication is to: (a) have defined and functioning channels of communication, (b) to make the flow of information work between operators, (c) to have clearly assigned responsibilities and competencies (e.g., who informs in which situation), and (d) to make sure that the receivers of the message know they can trust the information and that they are prepared to act (Lombardi, 2005 in Höppner et al., 2010). These are challenges that need to be addressed by the three CBSOs in collaborating and partnering to create a comprehensive local response to disasters.

ACKNOWLEDGMENTS

The authors gratefully acknowledge the cooperation of Bandung Planning and Development Agency, Bandung City Government and Bandung Women Associations in the central, subdistrict, and ward levels of Bandung, Indonesia. Advice and insight from the head of the Bandung City Women Welfare Associations are also highly acknowledged. The first author is also thankful to the Japanese Government (Monbukagakusho, MEXT) for the scholarship and the support received from the Global Center for Education-Human Security Engineering for Asian Megacities (GCOE-HSE) Program of Kyoto University in conducting this study.

REFERENCES

Adger, W. N. (2000). Social and ecological resilience: Are they related? *Progress in Human Geography*, *24*(3), 347–374.

CDASC. (2008). *Local initiative for global resilience*. Completion Report of 2008. Child Disaster Awareness for Schools and Communities (CDASC) Management as part of Muhammadiyah Disaster Management Centre, Central Board of Muhammadiyah, Jakarta, Indonesia.

Clayton, A., Oakley, P., & Taylor, J. (2000). Civil Society Organizations and Service Provision. Civil Society and Social Movements Program Paper No. 2. United Nations Research Institute for Social Development/UNRISD, Geneva, Switzerland.

Covello, V., & Sandman, P. M. (2001). Risk communication: Evolution and revolution. In A. Wolbarst (Ed.), *Solutions to an environment in peril* (pp. 164–178). Baltimore, MD: John Hopkins University Press.

Cvetkovich, G. (Ed.). (2000). *Social trust and the management of risk*. London: Earthscan Publishers.

Helsloot, I., & Ruitenberg, A. (2004). Citizen response to disasters: A survey of literature and some implications. *Journal of Contingencies and Crisis Management*, *12*(3), 98–111.

Höppner, C., Bründl, M., & Buchecker, M. (2010). *Risk communication and natural hazards*. CapHaz-Net WP5 Report, Swiss Federal Research Institute WSL. Retrieved from http://caphaz-net.org/outcomes-results/CapHaz-Net_WP5_Risk-Communication.pdf

IFRC. (2007). Indonesia: Jakarta floods. Information Bulletin No. 4/2007, Appeal No. MDRID003, September 26, 2007.

IRI, UNOCHA, & LPBI NU. (2011). *Preliminary study of the potential role of mosque in disaster situation in Indonesia: Six mosques in two provinces.* Islamic Relief Indonesia (IRI), United Nations Office for Coordination of Humanitarian Affairs (UNOCHA), and Climate Change and Disaster Management Institution of Nahdlatul Ulama, Jakarta, Indonesia.

Joerin, J., & Shaw, R. (2011). Mapping climate and disaster resilience in cities. In R. Shaw & A. Sharma (Eds.), *Climate and disaster resilience in cities* (Vol. 6, pp. 47–61). Community, Environment and Disaster Risk Management. Bingley, UK: Emerald Group Publishing Limited.

Kaldor, M. (2003). Civil society and accountability. *Journal of Human Development, 4*(1), 5–27.

Kamat, S. (2003). NGOs and the democracy: The false saviors of international development. *Harvard International Review, 25*(1), 65–69.

Kieffer, C. (1984). Citizen empowerment: A developmental perspective. *Prevention in Human Services, 3*, 9–36.

Leitmann, J. (2007). Cities and calamities: Learning from post-disaster response in Indonesia. *Journal of Urban Health: Bulletin of the New York Academy of Medicine, 84*(1), i144–153.

McEntire, D. A. (2001). Triggering agents, vulnerabilities and disaster reduction: Towards a holistic paradigm. *Disaster Prevention and Management, 10*(3), 189–196.

Ministry of Social Affairs. (2010). *Ministry Regulation No. 77/HUK/2012 of the Guidelines of Youth Union of Indonesia.* Court of Constitution of Republic Indonesia. (In Indonesian). Ministry of Justice of Republic Indonesia. Retrieved from http://portal.mahkamahkonstitusi.go.id/eLaw/mg58ufsc89hrsg/1ffcf804f279dc9761388fca540279705bbdaefd0.pdf

Ministry of Social Affairs. (2011a). *Pemberdayaan Karang Taruna dengan Program LPM, Ministry of Social Affairs of Republic Indonesia.* Retrieved from http://www.depsos.go.id/modules.php?name = News&file = article&sid = 355. (In Indonesian).

Ministry of Social Affairs. (2011b). *Taruna Siaga Bencana, Ministry of Social Affairs of Republic Indonesia.* Retrieved from http://tagana.depsos.go.id/?profile/history. (In Indonesian).

Mulyasari, F., & Shaw, R. (2012). Civil society organization and disaster risk reduction in Indonesia: Role of women, youth, and faith-based groups. In R. Shaw (Ed.), *Community-based disaster risk reduction* (Vol. 10, pp. 131–150). Community, Environment and Disaster Risk Management. Bingley, UK: Emerald Group Publishing Limited.

Mulyasari, F., Shaw, R., & Takeuchi, Y. (2011). Urban flood risk communication for cities. In R. Shaw & A. Sharma (Eds.), *Climate and disaster resilience in cities* (Vol. 6, pp. 225–259). Community, Environment and Disaster Risk Management. Bingley, UK: Emerald Group Publishing Limited.

Mulyasari, F., Shaw, R., & Takeuchi, Y. (2015). Women as disaster risk reduction drivers in bandung. Indonesia: Paving towards a resilient-community. *Disasters, 39*(3), in press.

National Research Council. (1989). *Improving risk communication.* Washington, DC: National Academy Press.

Paton, D., & Bishop, B. (1996). Disasters and communities: Promoting physiological well-being. In D. Paton & N. Long (Eds.), *Psychological aspects of disaster: Impact, coping, and intervention* (pp. 255–268). Palmerstone North: Dunmore Press.

Putnam, R. (2000). *Bowling alone: The collapse and revival of American community* (pp. 288–290). New York, NY: Simon and Schuster.

Shaw, R., Takeuchi, Y., & Mulyasari, F. (2012). Bandung Kecamatan Profile: Climate and disaster resilience. Kyoto University – Global Center of Excellence Program on Human Security Engineering for Asian Megacities, Bandung City Government, and Institut Teknologi Bandung, Kyoto, Japan.

Steinberg, F. (2007). Jakarta: Environmental problems and sustainability. *Journal of Habitat International*, *31*, 354–365.

TP PKK. (2006). PKK Cadre Handbook Year 2006, PKK Movement of West Java Province. [*Buku Saku Kader PKK. Tim Penggerak PKK Provinsi Jawa Barat Tahun*]. (In Indonesian).

Victoria, L. (2009). Community capacity and disaster resilience. In R. Shaw & R. R. Krishnamurthy (Eds.), *Disaster management: Global challenges and local solutions* (pp. 338–35). India: Universities Press.

CHAPTER 11

COMMUNICATING CLIMATE CHANGE AND ITS IMPACT TO DHAKA'S URBAN POOR – GRASSROOT WORKSHOPS FOR ADULTS AND CHILDREN

Annika Salingré and Boris Braun

ABSTRACT

Bangladesh's urban poor face numerous obstacles, especially during times of disastrous natural events. The effects of global warming will pose new threats for them as frequencies and magnitudes of natural hazards are likely to increase. Thus, current and formerly successful coping and adaptation strategies will be challenged in the future. The urban poor need to be informed and empowered so that they can develop their own strategies in response to the expected local effects of climate change. Our research aims at grassroot workshops as a means of communication with slum dwellers in Dhaka which not only convey possible climate change impact on livelihoods but also create relevance for the individuals and options for action. More than 20 workshops including men, women, and children have been conducted and analysed.

Risks and Conflicts: Local Responses to Natural Disasters
Community, Environment and Disaster Risk Management, Volume 14, 251–278

ISSN: 2040-7262/doi:10.1108/S2040-7262(2013)0000014017

The final workshop design incorporates a mix of participatory and input-oriented elements in order to overcome the problem of psychological distance.

Keywords: Communication; urban poor; natural hazard; grassroot workshop; Bangladesh

INTRODUCTION

Bangladesh is renowned for frequent natural disasters such as floods, cyclones, and earthquakes. Coupled with its relative poverty and low development, these impose a major challenge to the country and its people. Thus, coping with natural disasters is a common aspect of the lives of many Bangladeshis. Local responses and coping strategies are numerous and astonishingly successful. Even though, for example, dealing with flooding during rainy season is a usual aspect of traditional rural lifestyles, natural extreme events can still be highly threatening. Particularly the livelihoods of the poor come under pressure in these cases and it is those people who have to rely on meager resources and relatively simple local responses towards these threats. As traditionally people live in villages and indigenous responses have been developed in rural contexts, urban dwellers are facing new challenges and have to come up with innovative strategies.

Global warming is affecting regional climates. Significantly changing weather patterns have to be expected in the future and the magnitude of natural extreme events are likely to increase. Based on the results of previous research (for Bangladesh, see e.g., Aßheuer, Thiele-Eich, & Braun, 2013; Azam & Falk, 2013; Braun & Aßheuer, 2011; for parallel results for India, see Chatterjee, 2010), we can assume that there is an extensive need and potential for revised local responses. Lacking background knowledge and low levels of self-organization are the main obstacles to local responses by the urban poor. Thus, our project aims at knowledge sharing and community-based initiatives. Consequently, we developed workshops for and with slum dwellers in Dhaka that not only communicate possible climate change impact on livelihoods but also create relevance for the individuals and options for action. 21 workshops involving men, women, and children were conducted and analysed. The final workshop design incorporates a mix of participatory and input-oriented elements in order to overcome the problem of psychological distance.

DHAKA AS A HAZARD-PRONE AREA

Bangladesh's geography comprises a number of generally favorable conditions. It is characterized by an auspicious tropical climate – its morphology is shaped by many rivers and streams running towards the Bay of Bengal. As a result, Bangladesh's soil is mostly fertile. But at the same time, Bangladesh is one of the world's countries most prone to natural hazards (Beck, 2005). These include cyclones, droughts, earthquakes, floods, and river bank erosion (Agrawala, Ota, Ahmed, Smith, & van Aalst, 2003; Beck 2005; Braun & Shoeb, 2008; Edris & Collins, 2010; Hofer & Messerli, 2006; Novak, 1994; Reliefweb, 2007; Sen, 1981). According to Edris and Collins (2010), 97 percent of the country and 98 percent of the population are hazard-prone.

The extent of affectedness coupled with the variety and number of natural disasters leads to high numbers of fatalities (Paul & Routray, 2011). The most regularly occurring disaster is flooding which mainly includes rain floods and river floods as well as urban water logging. The frequency and magnitude of flood events in Bangladesh is based on its deltaic morphology with more than 60 percent of the land area below 6 m above sea level as well as the huge and rain-laden catchment areas of mighty rivers. As a result, floods are frequently experienced by the people of Bangladesh (Ahmed & Falk, 2008; Braun & Shoeb, 2008; Hofer & Messerli, 2006; Rashid, 1991).

Forecasts on regional effects of global warming are difficult to make and remain relatively vague due to the variety of influencing factors. Analyses of the data of the Bangladesh Water Development Board show that in the past 60 years the annual flooded area has not increased significantly in the long term (see Fig. 1). Due to the increased variability of rainfall and precipitation patterns, the variability of floods from year to year has however risen considerably (Islam & Neelim, 2010). A crucial factor for this development is the increasing global average temperature (Palmer & Räisänen, 2002). Further consequences of increasing temperatures are decreasing precipitation in winter, increasing precipitation during other seasons increasing likelihoods of extreme rainfall events, and flooding as well as the potential rise in sea level which may add to flooding by decreasing the gradients of the rivers (Ali, 1999; Christensen et al., 2007; Cruz et al., 2007; Hofer & Messerli, 2006; IPCC, 2012). While these natural conditions are becoming more and more critical, migration to urban agglomerations, especially Dhaka, continues to lead to growing population concentrations in disaster-prone locations (Walsham, 2010; UN, 2012).

The future effects of global warming will add further to the already high frequency and magnitude of natural disasters in Bangladesh and, thus, impose new threats on the country and its population. The capital Dhaka is a megacity of roughly 15 million inhabitants (UN, 2012) with about 40 percent of the population living in slum settlements which are spread all over the city. In these areas, the majority of Dhaka's growth and the highest numbers of urban poor are recorded (CUS et al., 2006). Most of them work in the informal sector. The average monthly income of these households is below TK 5,000 (approx. /US$ 59; see CUS et al., 2006). Thus, most of the slum dwellers face severe difficulty concerning their livelihoods and everyday survival. The environmental and infrastructural situation in slums is hazardous in many cases, creating a large number of social, hygiene, and health-related problems. As only very few public services and means of support are available for the slum dwellers, they manage their problems independently and have developed various coping and adaptation strategies. People apparently live "from hand to mouth" and somehow manage to survive. The advancing local effects of global warming will cause increased pressure on the livelihoods and adaptation capacity of the people as well as increased hazard-proneness. This affects people's ability to work, leads to cutbacks in nutrition and an aggravated health situation for the urban poor. Thus, vulnerability increases (Aßheuer et al., 2013). Accordingly, early information on climate change and improved knowledge on preparation is indispensable.

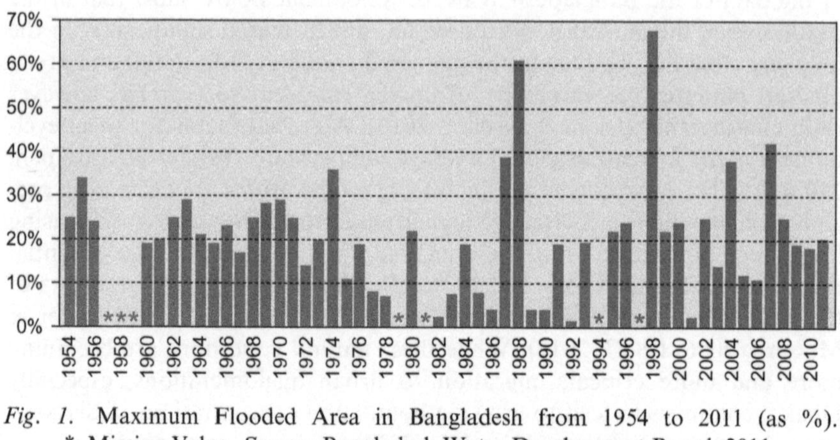

Fig. 1. Maximum Flooded Area in Bangladesh from 1954 to 2011 (as %). *: Missing Value. *Source:* Bangladesh Water Development Board, 2011.

Each of the more recent floods in Bangladesh also had severe effects on Dhaka (Alam & Rabbani, 2007; Islam, 2005; Stalenberg & Vrijling, 2009). Previous research on the one hand reveals that most households in slums are highly vulnerable due to poor financial, physical, and human resources but, on the other hand, also show a surprising capacity to cope with floods due to informal social networks (Aßheuer, Salingré, Thiele-Eich, Braun, & Simmer, 2012; Braun & Aßheuer, 2011). Even though social capital helps the slum dwellers to cope with, and to a certain extent, proactively adapt to floods, coping does not automatically mean long-term socioeconomic development or increasing resilience (Aßheuer et al., 2013; Braun & Aßheuer, 2011). Accordingly, improvements in options for action through knowledge and community organization are required to build up resilience. Based on the knowledge on current coping strategies, their capacity and limitations, as well as regional climate change scenarios, we aimed at the communication of scientific findings on floods, climate, and climate change to Dhaka's slum dwellers. The capacity to adapt and learn is relatively low, resulting in the poor getting by but rarely getting ahead. Building on this, we aimed at strategies, tools, and principles for the communication of climate change-related disaster risks and possible impacts on the livelihoods of the slum dwellers and to empower them to initiate a revision of current strategies and community-based adaptation with regard to natural hazards.

CHARACTERISTICS OF THE URBAN POOR IN DHAKA

In order to develop a suitable methodology, the target group and local conditions need to be analysed in detail. Our project focused on various informal and formal marginal settlements in Dhaka. Our study sites were chosen from the entity of settlements fulfilling the slum criteria by the local Centre for Urban Studies (CUS) (Table 1).

Despite the heterogeneity of slum settlements (e.g., with respect to legal status, size, and location) and differences between individual households, the slum dwellers generally share a number of characteristics which are relevant in risk communication: the adults are usually familiar with natural hazards in general, but due to the fact that most families are migrants from rural areas, many of them do not have any experience at their current place of residence. Hence, the use of previous experience and existing knowledge are limited when targeting new challenges. Furthermore, the migratory background also affects the long-term perspectives of many households.

Table 1. Essential Characteristics of Urban Slums in Bangladesh According to CUS.

More than 50% of the dwellings are in very poor conditions, e.g. shacks made of temporary materials.
High population density and/or crowding of rooms, e.g. average densities above 75,000 inhabitants per km^2 and/or less than $4m^2$ living space per person.
Poor supply and sanitation, especially with regard to hygienic conditions and water.
Low socioeconomic status of population, accordingly at least 50% of households with monthly incomes of TK 5,000 or less.
Insecurity of tenure.

Source: CUS et al. (2006).

Since the people originally came to Dhaka for economic reasons, they still hold strong ties with their families back in the villages. Thus, intentions for remigration are widespread. In many cases, people may finally stay in Dhaka, but as long as they plan to leave again in the foreseeable future, long-term strategies for development and improvement at the individual level are hampered. General education is widely missing as especially adults lack formal schooling (literacy levels are below 50 percent amongst household heads according to Braun & Aßheuer, 2011). In contrast, however, the slum dwellers have valuable indigenous local knowledge of relevance regarding rural lifestyles and environments. Finally, the precarious living conditions of the people call for a prioritization of daily survival. Accordingly, it is often difficult for them to allocate time and resources to attend the workshops and to continue to deal with the respective issues afterwards. These problems are even reinforced by the powerlessness of the slum dwellers who often find themselves at the mercy of other, more influential people (local politicians, bureaucrats, NGOs, etc.).

Despite and particularly because of the limited options and resources for community-based actions, our workshops for empowerment had to consider and address the special conditions and challenges described.

DIDACTIC PRINCIPLES

In order to address the aforementioned challenges in risk communication, several didactic principles proved to be crucial and were incorporated into the workshop design. Knowledge is always a combination of learning and experience. This means that linking functional knowledge, life experience, and emotional experience can lead to a better understanding of complex issues (Bonifacio, Takeuchi, & Shaw, 2010; Gropengießer, 2008; Schüßler, 2004). Accordingly, it is crucial to embed new information into a network of already known information and indigenous knowledge. This phenomenon is called *networked knowledge* and has been one of our core principles.

Another crucial factor for leaning is *motivation*. The more eager the participants are to learn, the more energy, enthusiasm, and concentration will be put into the situation from their side. Thus, participants need to know why they are supposed to learn about a specific issue, know about the relevance of and the benefit from learning, and should actively be motivated to participate by the trainers or workshop leaders.

Climate change is an abstract and psychologically distant issue as it goes beyond normal sensual perception. The more distant a topic is, the less people think about it and, thus, it appears to be less relevant (Beyerl, 2010). By application of the *relevance* principle "distance [can] be reduced by making climate change more real, local, relevant and immediate to the people which might promote action on climate change" (Spence, Poortinga, & Pidgeon, 2011, p. 1). The incorporation of local examples and incidents from everyday life which happen in the participants' surroundings allows for increased emotional and cognitive engagement. Therefore, it is important to focus on specific aspects in order to address abstract and psychologically distant topics and ease understanding (O'Kane, 2008; Schüßler, 2004). This approach is based on the finding that clearly defined and detailed goals as well as the psychological proximity of a topic significantly increase the likelihood of action (Spence et al., 2011).

Even though relevance often goes along with personal interest or concern, it does not automatically imply the *need and options for action*. In order to ensure that people memorize the information delivered in the workshop and even more to initiate potential activities, the relevance of action for the participants had to be clarified. This aspect was realized by discussing currently applied strategies on the basis of potential changes. At many points, the participants discovered that the strategies might not be functional in the future and thus revised and innovative strategies are needed. Still, this has to be illustrated considerably. It is essential not to create an impression of current strategies becoming malfunctional. Instead, the relevance for action has to be a starting point for discussing, rethinking, and developing (new) strategies. It is important to create realistic options for actions in the workshop, provide the participants with the means to work on these, and give them the motivation and strength to enter this process.

Finally, a *mix of methods and media* is helpful to keep the workshops interesting and enable the participants to pay attention more easily. The methods used have to comprise input-oriented as well as participatory elements, they should help to highlight the main information, and address cognitive and affective dimensions of learning. Media should be used in order to illustrate the issues discussed but not lead to a visual, auditory, or any other overload of information and impressions or distract the participants. Furthermore, the media need to be adequate. In the context of our workshops that meant bearing in mind the participants' age, gender, and life experience; the widespread illiteracy; as well as potential cross-cultural differences especially with regard to symbols and metaphors used.

METHODOLOGY: WORKSHOP DEVELOPMENT, EVALUATION, AND ANALYSIS

In preparation for the workshops, an extensive literature review was carried out in order to learn about other projects' methods and results in conveying expert information to the local population and awareness-creation workshops. All the approaches that have been developed so far were rooted in various forms of participatory appraisals, such as participatory rural appraisal (PRA) (see e.g., Absalom et al., 1994; Chambers, 1996) and participatory learning and action (PLA) (see e.g., Pretty, Guijt, Thompson, & Scoones, 1995; Schönhuth, 1996; Thomas, 2002). However, it was essential for our project to adapt these methods and principles to the given topic of weather extremes and climate change and also to the specific social, cultural, and natural environment of slums in Dhaka. In order to root our methods in local communities, we cooperated with international and local partner NGOs. The local NGO *Glory Friendship Social Welfare* (GFSW), in collaboration with *Ärzte für die Dritte Welt* (German Doctors), runs health services as well as schools for slum dwellers. Additionally, scholars from the *University of Rajshahi* as well as Bangladeshi education experts from the Dhaka-based NGO *Interaction Bangladesh* cooperated in the project.

All the workshops were run by a German–Bengali team, in which the German team members played the role of experts whereas the Bengali team members played the role of facilitators and interpreters, according to a fixed workshop guideline based on the aforementioned didactic principles. The language of the workshops was Bangla and the phases held by the

German team members were simultaneously translated to the participants. The workshops took place in schools, public places, and shacks in six different informal settlements (Hazaribag, Mirpur, Box Culvert, Manda, Genderia, and Korail) throughout Dhaka (see Fig. 2). At each place, separate sessions with men, women, and children were held. The workshop concepts for men and women were almost similar, whereas the workshop layout for children differed significantly (see Table 2). The majority of our workshops for children were carried out with classes from GFSW slum schools. In all adjacent slum communities (Manda, Genderia, Korail), workshops with adults took place and were organized with the help of GFSW. Furthermore, the GFSW staff served as consultants and took part in teacher training on the scientific foundations of climate change. Consequently, this cooperation was a means of embedding the issue in educational institutions as well as in the communities.

In two field phases in 2011 and 2012 a total number of 21 workshops were conducted. All the workshops were videotaped, translated, and transcribed. After each workshop, feedback sessions and discussions among the project team members were held in order to analyze and improve the workshop concept. Additionally, the taped workshops were analysed after the completion of each field phase and systematic evaluation was carried out. In this way, subsequent revisions and improvements were possible.

Our evaluation aimed at analyzing the positive and negative aspects of each workshop as well as their dynamics in order to improve the concept as well as the facilitation. While it was possible to draw inferences about the immediate understanding and learning of the participants, long-term evaluation of the success of the workshops, that is, the performance of the participants in a given future situation, was not feasible within the scope of the project. The audiovisual data of the workshops allowed for reflection on action and illustrative proofs. Still, even though audiovisual analysis has been used in social research for several years integrative methods for analysis are still lacking (Rosenstein, 2000; Schnettler & Raab, 2008; Schubert, 2002). The audiovisual data was analysed together with the translated workshop transcripts. The essential steps of qualitative content analysis according to Mayring et al. (Mayring, 2000; Mayring, Gläser-Zikuda, & Ziegelbauer, 2005) were carried out with help of the software atlas.ti. Codes used for analysis were clustered in two groups: The first one was *motivation/interest* which included *active participation, topic-related enquiries, attention*, and *addressing other topics*. The second group was *training success* which included *previous knowledge, link to personal environment (of participants), realisation of relevance of climate change, comprehension of*

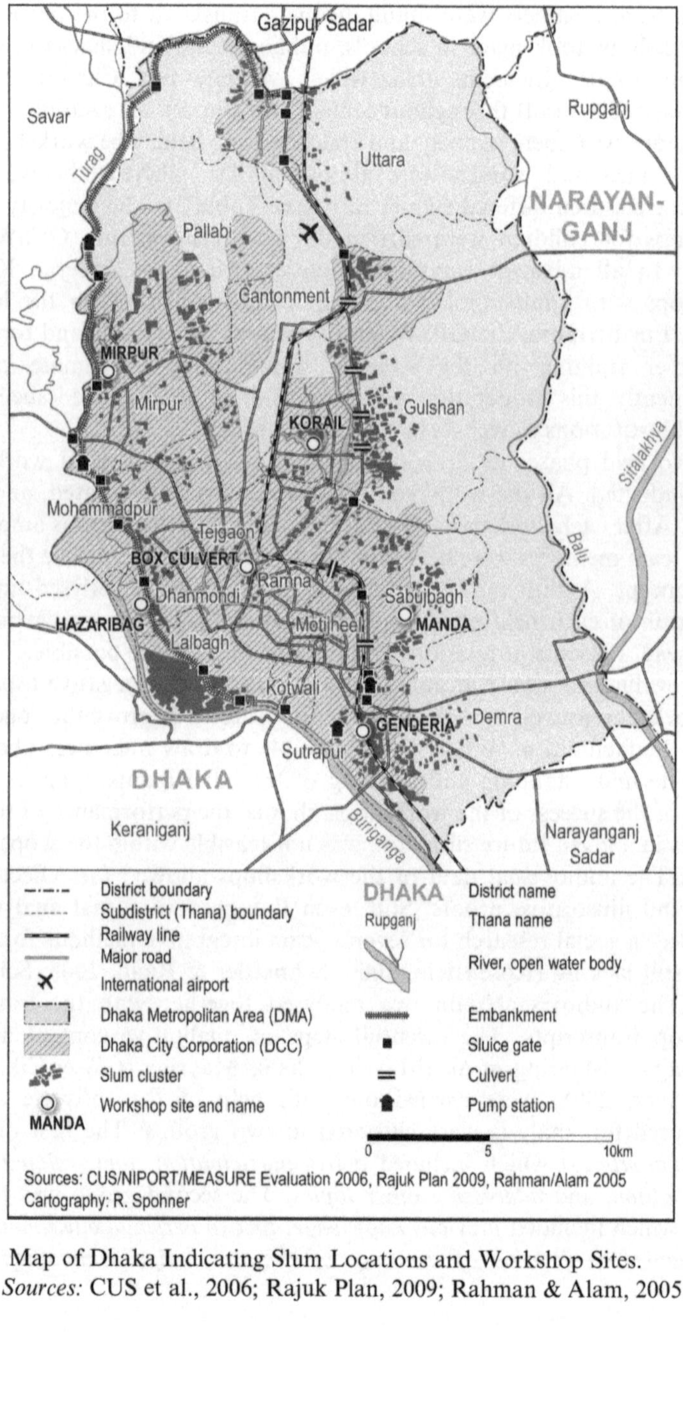

Fig. 2. Map of Dhaka Indicating Slum Locations and Workshop Sites.
Sources: CUS et al., 2006; Rajuk Plan, 2009; Rahman & Alam, 2005.

Table 2. Framework for Workshops.

Children	Adults
Approx. 2hrs during teaching time	Approx. 1 hr in the afternoon or early evening
Retrospective focus on flooding	Retrospective focus on climate-related hazards
Discussion of possible future strategies	Discussion of possible future strategies

contents, and *problem solving.* Additionally, *further influences* on the workshop situation were analysed. Amongst them *mistakes* (as compared to the guideline) by the workshop team, *external disturbances, persons leaving/entering the workshop, group dynamics, controlled/uncontrolled situations,* and *scepticism towards the workshop/workshop team* were taken into account

THE WORKSHOP CONCEPT

The didactic aspects illustrated above are crucial for the conceptual background of our work. Our workshops, thus, integrated the existing knowledge of the participants and included an introduction building on the people's experience of the different seasons and discussing their perceptions of changing weather patterns in the last few decades. The presentation of climate change and its effects focused on the local level and on noticeable changes. It was followed by a discussion of the impact of climate change on people's lives and their adaptation and coping strategies to illustrate the relevance for them and to reduce psychological distance.

As the understanding of the concept of climate change requires a comprehension of the concept of the climate as opposed to other phenomena, the logical sequence of our workshops started out as a very basic one and proceeded in small steps. First, the term and the concept of weather had to be clarified. This was illustrated with help of the local range of weather phenomena generally observed by the participants. Then, different kinds of weather phenomena were allocated to the different seasons and these were revised with the help of season cards depicting typical images such as the harvesting of certain crops or children playing in a heavy downpour. In Bangladesh, the course of the year is traditionally divided into six seasons: summer "grishshokal" which is very hot, with a lot of sunshine and many thunderstorms as well as some rain (March–May); the rainy season "borshakal" which has a lot of rain and many clouds in the sky (June–August);

autumn "shorotkal" which does not have much rain and sometimes the nights may be cool (September–October); late autumn "hemontokal" which is dry with generally cold nights (October–November); winter "shitkal" which is cold and foggy, especially in the mornings (November–January); and finally spring "boshontokal" which is when it starts to become hot again and the flowers blossom (January–February).

Taking up these seasons, we discussed the annual cycle of typical weather patterns during the course of the year which has been following the same pattern for a very long time. It was crucial to clarify that changing weather during the year is due to the seasons and is very typical and normal. So this does not require the term "change" and does not get our special attention as it is a natural and balanced process. Thus, we defined climate as a long-term "normal" repetitive cycle of weather phenomena including some extreme events occurring every now and then. By contrast, climate change means significant noticeable long-term changes to the climate.

Adults' Workshop

After a general introduction to the workshop situation, the team, the procedures of translation and videotaping, the benefits of participation, and the opportunity to ask questions to the team, a personal introduction took place. All the team members and workshop participants introduced themselves, stating their names and occupations. On the one hand, this introduction served as an icebreaker and helped to familiarize the participants with talking in front of the group and listening to each other. On the other hand, the occupations of the participants and their self-presentations also allowed some assumptions about the social setting of the slums and possible hierarchies within the group. First of all, after the warm-up phase, the assessment of previous knowledge took place. It included a very open question about associations to climate change and, subsequently, the question whether the participants thought it is something positive or negative. The first phase of the workshop was related to the normal climatic situation in Bangladesh. In this phase, the six seasons of Bangladesh were repeated or introduced with the help of visualizations and descriptions of typical weather patterns, depending on the participants' previous knowledge. In the course of this phase, the typical challenges and problems during the rainy season and winter were discussed in the group. Until this point, the workshops were held in a participatory manner by a Bangladeshi team member. To highlight the importance of the core information, only then was the topic of climate change taken up and explained with the help of specific inputs by the German team member who fulfilled the role of a climate change expert (the second phase). Climate change was then contrasted with seasonal weather change. Moreover, it was explained based on long-term trends which started in the past. The input only briefly covered the causes of climate change and also omitted effects which are not of local relevance in order to keep things simple and relevant to the given setting. After this input, the participants were given time to ask questions on climate change. In some cases, they took the opportunity and asked for repeated explanations or had in-depth questions. In order to assess whether the participants understood the term "climate change" and to enhance learning by repetition, the participants were encouraged to explain the term in their own words. The third phase of the workshop focused on the impact on adaptation and coping strategies and the revision and development of possible strategies. Based on the memories of the participants, a chart of floods between the 1980s and 2005 was drawn (cf. Fig. 3) and compared to the "official" chart of flooding in Bangladesh as depicted in Fig. 1. It was noticeable that the charts in all the workshops depicted the same extreme events (Fig. 3).

Based on the expected increase in floods, group work and discussions on the suitability of current strategies and possible new strategies took place (Fig. 4). The closing phase served to wrap up the workshop content as well as to highlight the relevance for action and the communities' potential once again (Table 3).

Children's Workshops

As today's youth will be more affected by climate change than the generation of adults and pupils are more used to learning new content through teaching situations, we decided not only to address adults but also children. At each study site, at least one group of pupils was chosen to attend a workshop which was conducted in their classroom during regular class time. The children usually were the first slum dwellers to be addressed as they were supposed to tell their parents about what they had experienced at school and, thus, raise their parents' interest in our work. Furthermore, the cooperation with slum schools offered an institutional link to the communities and enabled us to introduce the topic and new methods of instruction

Fig. 3. Sketch of a Flood Magnitude Diagram as Drawn by Participants in the Workshops.

Fig. 4. Scene from the Women's Workshop in Hazaribag. (Photograph: Nils Jahn, 2011).

within the schools. Most workshops were conducted with students aged 8 to 11 as children in slums often drop out of school after class four.

The children's workshops were based on the same theoretical background and didactic principles as the workshops for adults (Fig. 5).

Table 3. Guideline for Adults' Workshops (Simplified and Shortened Version).

1	Getting started: welcome, workshop introduction by Bangladeshi team member	BTM
2	Getting to know each other, breaking the ice	BTM
3	Overview of previous knowledge: asking about knowledge and associations regarding the term "climate change"	BTM
4	Examples and life experience of rainy and dry seasons with references to the study site and home villages of the participants with the help of illustrative photographs	BTM
5	Participatory introduction of six seasons with the help of season cards	GTM/BTM
6	Explanation of the term "climate" with reference to the seasons	BTM
7	Input on climate change: explanation of the term "climate change" and its causes with help of pictograms	GTM/BTM
8	Review: explaining climate change in the participants' own words	BTM
9	Group work finding solutions	BTM
10	Thanks and farewell	GTM/BTM

However, as the target group and workshop setting differed significantly from the workshops with men and women, the guideline was adjusted in many ways. Due to the availability of a blackboard and the literacy of the children, blackboard charts, reading, and writing exercises were included (Fig. 6). As the children were familiar with different forms of teaching and learning, we also made use of group work, small presentations, games, and creative tasks (Table 4).

Implementation of Didactic Principles

To establish *networked knowledge*, a lot of emphasis was laid on the linkages between previous and newly gained knowledge. In the first part of the workshops, an assessment of previous knowledge was carried out. In some cases, participants had more or less clear ideas about climate change and its impact, others had incorrect knowledge. Most of the participants had no knowledge about the topic at all. Based on the assessment, the seasons of Bangladesh were introduced or repeated, the terms "weather change" and "climate" were introduced and demarcated. Only after these steps, was the term climate change introduced and explained. Past long-term trends and expected future changes were mirrored with people's personal perception.

Fig. 5. Scene from the Children's Workshop in Hazaribag (Photograph: Nils Jahn, 2011).

Fig. 6. Exemplary Blackboard Image Used During a Children's Workshop in Manda. (Photograph: Annika Salingré, 2012).

In order to *create relevance* for the slum dwellers, all the information was directly and explicitly applied to their living situation. Discussions focused on their problems in everyday life during times of natural disasters and in relation to prospective future developments.

As the workshop team could not provide distinct strategies in response to the local effects of climate change, the goal was to initiate a problem-solving

Table 4. Guideline for Children's Workshops (Simplified and Shortened Version).

1	Getting started: welcome, introduction of guests by the teacher, workshop introduction	BTM
2	Getting to know each other, breaking the ice	BTM
3	Dream Journey	BTM
4	Overview of previous knowledge: asking about knowledge and associations regarding the term "climate change"	BTM
5	Input on and participation in the term "climate": introduction of six seasons with the help of season cards, explanation of the term "climate" with reference to the seasons	GTM/ BTM
6	Example of the climate change at the study site: comparison of personal experience of local weather conditions and natural extreme events and future scenarios with the help of photographs	BTM
7	Input on climate change: explanation of the term "climate change" and its causes with the help of pictograms	GTM/ BTM
8	Review: writing exercise explaining climate change in the puplis' own words	BTM
9	Group work finding solutions	BTM
10	Drawing exercise: changing situation during climate change	BTM
11	Thanks and farewell	GTM/ BTM

process and to empower the communities. In order to encourage independent community processes, the workshops included participatory elements and initiated exchange amongst the slum dwellers themselves. Participation also strengthened their self-esteem and helped to keep their concentration high throughout the whole workshop duration.

Motivation, in general, was an important issue to take into consideration for the design and conduct of the workshops. On the one hand, the urban poor are very busy with income generation and household work and cannot afford a lot of time for participation in workshops. On the other hand, based on their previous experience, they expect foreigners to deliver direct help and are somewhat surprised if they are not directly given material assistance or explicit advice. The goals of knowledge sharing and exchange, therefore, had to be transparent from the beginning of the workshops and the participants needed to be encouraged to stay for the whole duration of the sessions as well as to be actively motivated by the workshop team with the help of several activities.

The *need for action* was illustrated by discussing current strategies in the context of potential future weather patterns and natural hazards. It became obvious to the participants that the potential future situation will very likely challenge their current coping and adaptation strategies and that a revision of the latter will be needed. However, this insight might leave people desperate as they are facing an increasingly problematic situation and are barely getting by even at the moment. Thus, the identification of *options for action* was crucial. In plenary and small group discussions, the current strategies were tested and the potential for adjustments was thought through. As it is, of course, impossible to thoroughly review the actions of the communities in no longer than an hour, we focused on initiating the process of exchanging and discussion of strategies within the communities, resulting in closely assessing and carefully improving these. Ideally, the participants left the workshops with an awareness of the need for the development of sustainable local responses as well as trust in their communities as capable drivers for this process.

The methodology of the workshops was shaped by a mix of input-oriented and participatory *methods*. Especially the linking of new and existent knowledge as well as the development of options for action, the overview of local weather patterns, and the discussion on local seasons and climate were carried out in plenary discussions moderated and visually supported by the team. Actually, the only real input was the presentation of the concepts of climate and climate change and, furthermore, the potential local developments in the middle of the workshop. The *media* used mainly consisted of color-printed large photographs of typical scenes from the six seasons and from incidents of severe floods. For the adults' workshops, we additionally used the chart of a flooded area (Fig. 1) and a blank diagram in order to get an overview of the perceived and actual extent of floods (Fig. 3). In the children's workshops, icons for the illustration of the causes and effects of climate change were used in a blackboard image (Fig. 6) and in the course of some exercises the children wrote keywords and short texts and drew images on the blackboard.

LESSONS LEARNED

When working with the urban poor, one has to bear in mind that the target group is under multiple pressures. The adult slum dwellers are concerned about their livelihoods nearly all of the time. Men work in various occupations for many hours per day to generate an income, women also partly work outside their homes and, in addition, have to do the household chores and take care of the children. Thus, these people can scarcely allocate the time and energy to attend a workshop. Consequently, and also because the adults are not used to sitting, talking, and staying focused on a topic for a long time, the workshop has to be kept short. Breaks can be helpful under these conditions but, at the same time, cause a lot of disturbance and fluctuation. So finally, we ended up with nonstop workshops of approximately one hour (two hours for children). Still, this was too much time for many participants. Some left early, opted out, or fell asleep. However, we could not shorten the workshops any further as the concepts had to be introduced carefully, repetition was needed, and it was crucial to initiate certain processes.

Not only did time constraints force us to keep things simple, but the amount of new information for the participants was huge and it was important to avoid an overload. This would potentially have led to confusion and hampered the intended workshop outcomes. Thus, we decided to narrow things down and focus on crucial aspects. Our selected content included aspects from the participants' previous knowledge in order to build up links. Beyond this, the potential local effects of climate change were presented. Global-level effects and for the adults' workshops also the causes of global warming were purposely omitted as they are of minor relevance for the target group. With the children, we discussed the causes of climate change as some of them had already heard about these in school or from the media and it was important to embed as much previous knowledge as possible. They were also more familiar with classroom situations and focusing on a topic and, in many cases, had relevant basic knowledge in science and there were less time constraints as the workshops went on for about two hours. Simplification also played a role in the choice of terminology and presentation of causalities. In line with Moser and Stein (2011) we used the increasing frequency and magnitude of natural extreme events as proof of climate change as these are tangible manifestations of the vulnerability of the urban poor.

In general, participation in the workshops was easier for children than for adults. In the same way, it was less challenging to design and run the children's than the adults' workshops. This was due to a number of reasons: first of all, the children's workshops took place in schools during general teaching time. The pupils did not only prove to be generally enthusiastic but they were also happy about some alternation in their school routines. The foreigners coming to their class, the interesting topic, the uncommon methods used, the presence of a camera and a microphone, etc. made our workshop an exciting experience. Furthermore, the children were in their usual class settings and used to interacting in their groups and with teaching personnel. Thus, they had no difficulty in asking questions, expressing ideas, and discussing within their groups. The longer timeframe also allowed for a slower and rather playful progression in the workshops as well as some fun activities. Finally, it was helpful to have meetings with teachers and headmasters/headmistresses at the schools in order to make preparatory arrangements, adapt our concept to the given situation, and establish links to the communities of adjacent slum settlements.

The establishment of links to local communities was essential at all sites. Prior to the first workshops, we went to each site to meet key people from the community, to enable the dwellers to get to know us in advance, to decide whether workshops are feasible, and to decide and announce the time and place. In addition to schools, we tried to establish links to NGOs, community institutions, health centers, social workers, and local landlords. These organizations and individuals were able to provide basic information about the communities, their general problems, past incidents of floods, suitable workshop times and venues, and often offered to inform the members of the community. For us, it was crucial to have them onside and involve them in order to embed the new topic and initiated processes in the communities.

Generally, the concepts of climate and climate change as long-term phenomena are not easy to understand. The distinction between weather change in the course of a year and long-term changes in climate (global warming and its effects) were, thus, not easy to distinguish. In order to illustrate the timeframe for long-term changes, we referred to the seasons and to the climate one or two generations back. Adults were asked about their memories of seasonal change during their childhoods and whether they had ever heard accounts of elders talking about the weather in the past. In most cases, people were able to recount the same patterns of seasons as nowadays but soon started to list abnormalities they had noticed recently:

> "Oh! These six seasons [normal seasons of Bangladesh as discussed and presented in the workshop] are outdated."
>
> "The six seasons are increasingly changing."
>
> (Participants from the men's workshop in Mirpur)

These discussions were used to illustrate long-term patterns that are stable despite sporadic divergent and extreme events. However, after thorough and stepwise explanations, repetition and answering of questions, the participants were well able to understand about climate change, its potential local effects, and impact on their lives.

"We understood the issue of climate. That means, for example, that during the month of Kartika [October/November] it rains a little and nights are a little cold. Ok? Then it feels hot during the months of Boishakha [April/May] and Joishtho [May/June]. Again it feels very cold in winter."

"He is right, different types of weather in different seasons. From those, we get rains in rainy season, cold in winter."

"More heat during summer and more cold during winter. More winter and more warm. That is called climate change."

"Because of climate change, we might be facing more floods? That means less in some years and more in some years, but we have to face these."

(Participants from the men's workshop in Box Culvert)

Again and again, the workshop situations were shaped by the everyday problems of the slum communities. The people are constantly concerned with their current and apparently urgent problems and, therefore, do not have or do not allocate resources for dealing with climate change. That topic is psychologically distant while the more urgent problems of slum dwellers appear to be closer.

"Now the main problem is that if we get the electricity facilities, the owner of these houses will demand more rent for their house. It will be very difficult to stay here if we fulfill their demand. We don't have enough financial abilities to fulfill their demand."

"I am talking about the environment. I invested 17 lakh Taka. The place we live here is not hygienic for our health. The dust, the dirt, and stains are everywhere. That's why different kinds of diseases are increasing."

(Participants from the women's workshop in Hazaribag)

Spence et al. (2011) have analysed the subjective distance of climate change and its impact on the concern and actions of individuals which is highly relevant for risk communication and information on climate change effects in various contexts. Their analysis incorporates various dimensions such as spatial, temporal, and social distance as well as uncertainty. They found that the risks related to climate change are generally perceived to be far away in all four dimensions. High psychological distance correlates with low levels of concern, while low levels of concern correlate with little readiness for individual action. Consequently, in order to inform and empower the affected population so that they can revise their strategies, psychological distance has to be reduced as much as possible. Furthermore,

Spence et al. stress the fact that highly abstract and de-contextualized events tend to be psychologically distant. This leads to the suggestion to decrease the psychological distance by presenting climate change in a tangible manner, linking its effects to the context of the target group, and framing it with the previous knowledge of the participants.

Furthermore, the workshops can only be successful if they are run by a well-attuned team which consists of experienced, knowledgeable, and highly committed members. It proved to be very useful to inform all the team members extensively. We not only explained and discussed the background and aim of our project in general and the workshops in specific but carried out preliminary exercises as well as training workshops which were evaluated. Bangladeshi team members were much more than just interpreters. Their local and cultural knowledge as well as their evaluation of the workshops were highly beneficial for the whole research process. The more they got involved, the more committed they became. However, one cannot overemphasize the cross-cultural component of working in an international team. Things might apparently go smoothly but actually not be that way. This needs to be borne in mind throughout the cooperation, especially in the initial and training phases. In our cooperation, we had the experience that asking questions and making critical remarks was difficult for all team members for a certain time and we had to go through a couple of training and preparation loops until everything relevant was asked, expressed, and understood. In this stage, team leaders should include being open to learning in their role.

Finally, data analysis revealed and validated a range of insights which are of relevance for conducting workshops and employing other communication techniques: a thorough workshop introduction proved to pay off. Even though we discussed with community members whether they were interested in learning about climate change and discussing community-based strategies amongst themselves as well as explicitly announced information and discussion without financial or material support, there were many false expectations. At the same time, some community members remained sceptical about whether to trust us. Thus, the workshop introduction was revised and extended stepwise and presented to the participants including every detail. This was of special importance as participants did not usually ask questions about the workshop and the team in the beginning. These had to be anticipated and answered in advance. Analysis showed that unspoken and unanswered questions impacted the workshop atmosphere and participants' attention negatively. Consistent and well-structured workshop moderation/facilitation significantly coincides with the distinct active participation, motivation, and focus of participants whereas in cases of unsystematic workshop conduct, the workshops could not be carried out successfully with regard to the aspects mentioned.

Due to the language barrier, workshop moderation was very much in the hands of facilitators, and so was the consistency of leadership. However, local facilitators could be supported with regard to workshop flow and structure by the German experts. This led to the extended involvement of the experts in the moderation of some parts of the workshop and in providing assistance to the facilitators for some other parts and giving inputs. The time required for translation did not have any negative effects on the workshops. The participants were curious to listen to the strangers and behaved very patiently while waiting for translations. The translation process also slowed down the workshops which allowed time to think about the information provided, digest it, and phrase questions. However, retranslation and coordination within the workshops remained challenging even for the well-practised team. For well-facilitated workshops, the data clearly showed that external disturbances, such as people asking for items from the shacks, children crying, outsiders looking on, were disregarded by the participants or could easily be absorbed by the team. The same applied to people leaving or entering the workshop site and side discussions. It is clear from the audiovisual material that independent of the workshop quality, at each workshop there were participants who were highly enthusiastic, interested in the topic, and committed to participation. In the same way, we came across a few people who did not seem to be motivated and willing to take part at all, even though participation was supposed to be voluntary. At the end of the day, practical experience and data analysis reinforced the assertion that the quality and success of a method, in our case the workshops, is not only about concept, content, and preparation but a lot depends on adequate facilitation and translation. Team training and attunement proved to be essential.

CONCLUSIONS

Even though falling back on a well-established set of tools from participatory appraisals, methods for the assessment of local perceptions and strategies towards severe weather events as well as the initiation of community-based adaptation processes are still in their early stages (Moser & Stein, 2011). As the core of participatory methods were developed in a very hands-on manner in the 1990s (Absalom et al., 1994; Chambers, 1996), recent projects have mostly been concerned with questions on the situation and environment of their subjects of research. In contrast, we were concerned with methodological questions regarding result communication. Experience in this field is still very limited and each project is of an exploratory character. Moser and Stein (2011) have pointed at the unique potential of participatory methods such as the capability of the urban poor to identify the impact of global warming on their immediate environment and their daily lives. At the same time, results from Spence et al. (2011) highlight a core challenge in climate change communication: in various contexts, climate change and related issues were of lower priority compared to other different topics and perceived as temporally and spatially distant. Thus, this psychological distance had to be anticipated and overcome.

The integration and active involvement of the objects of research in the scientific process turns them into subjects as it is the core principle of participatory appraisals. This is not only a first step to acknowledge and integrate indigenous knowledge and local expertise but can also serve as a means to empower the local people and civic institutions and to make them agents of change (Fig. 7).

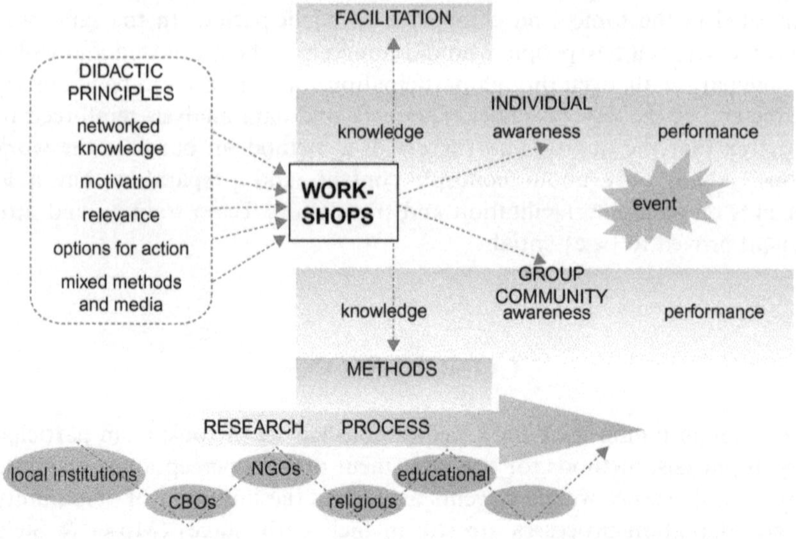

Fig. 7. Conceptual Diagram: Different Levels of Evaluation. *Source*: Author.

Despite the encouraging results and the positive progression of workshops, some general concerns and challenges remain. These might well apply to the communication of scientific results in other settings as well. A major concern is systematically testing and evaluating the effectiveness of the workshops. Success can be measured on different levels: (1) on the level of facilitation, (2) on the methodological level, (3) on the individual level, (4) on the group or community level or (5) with regard to the research process (see Fig. 7). First insights regarding levels 1 to 4 could be gained during and right after the workshops and could be wrapped up in immediate evaluation. Testing the knowledge gained was implemented in the workshop concept by assessing the participants' previous knowledge and comparing it to their own explanations of climate change at a later stage of the workshop. This was mainly carried out in order to find out whether further explanations and repetitions were needed or whether the concepts presented were understood by the participants. However, a systematic assessment of the knowledge gained generally also needs to be carried out with a certain time lag. The same holds true for evaluating the change in awareness.

The aim of initiating the acquisition of new competences and improving the performance in the case of a natural extreme event could not be tested in the scope of the project, as it required a relevant incident and the incorporation of a reference group. As the knowledge, awareness, and competence gained can hardly be measured directly after the workshop, it is even more desirable to cooperate with local partners and institutions that can facilitate an ongoing community process. For example, training local teachers can have a sustainable impact on teaching content and quality and, thus, lead to the improved knowledge and awareness of pupils. Furthermore, community-based organizations and local NGOs can carry on work on potential future strategies and moderate local discussion processes, while researchers usually quit the field after completion of their project. It is, thus, desirable to anchor recommendations and strategies from the workshops locally, both at the administrative level and in local and community-based organizations (Moser & Stein, 2011).

The participatory approach focuses on peoples' skills and capabilities and intends to sensitize and to trigger processes early on so that small-scale, modest, and incremental activities can be developed within communities to build up resilience against and to respond to recurrent patterns of severe weather impact. Since there is often limited access to municipal agencies for the urban poor, local NGOs might be the preferred local partners.

ACKNOWLEDGEMENTS

The authors thank the German Research Foundation for the financial support received for our research project "Natural hazards and climate change in Dhaka: future trends, social adaptation and informal dynamics" within the Priority Programme 1233 "Megacities – Megachallenge: Informal Dynamics of Global Change". Furthermore, the authors are grateful to Tibor Aßheuer who carried out the first stage of empirical research in the project. The authors also thank Clemens Simmer and Insa Thiele-Eich (University of Bonn) and AZM Shoeb and Raquib Ahmed (University of Rajshahi) for their kind cooperation. Finally, the fieldwork would not have been possible without the deep commitment of the interpreters, Nishat Jahan Jyoti, Anita Shafiq, and Laila Nur Shemonto, German students Magdalena Oppitz and Nils Jahn, and the kind support of Glory Future Social Friendship (Dhaka), Ärzte für die Dritte Welt (Frankfurt), Psychologen über Grenzen (Potsdam), and essentially the patience, hospitality, trust, and knowledge of Dhaka's slum dwellers.

REFERENCES

Absalom, E., Chambers, R., Francis, S., Gueye, B., Guijt, I., Joseph, S., ... Welbourn, A. (1994). Sharing our concerns and looking to the future. *PLA Notes*, *5*, 5–10.

Agrawala, S., Ota, T., Ahmed, A. U., Smith, J., & van Aalst, M. (2003). *Development and climate change in Bangladesh: Focus on coastal flooding and the Sundarbans*. Paris: OECD.

Ahmed, R., & Falk, G. (2008). Bangladesh: Environment under pressure. *Geographische Rundschau International Edition*, *4*(1), 12–18.

Alam, M., & Rabbani, G. (2007). Vulnerabilities and responses to climate change for Dhaka. *Environment and Urbanization*, *19*(1), 81–97.

Ali, A. (1999). Climate change impacts and adaptation assessment in Bangladesh. *Climate Research*, *12*, 109–116.

Aßheuer, T., Salingré, A., Thiele-Eich, I., Braun, B., & Simmer, C. (2012). Natural hazards and climate change in Dhaka: Future trends, social adaptation and informal dynamics. In M. M. Islam & W.-P. Zingel (Eds.), *Climate change effects and energy development* (pp. 132–136). Germany, Bonn: Bangladesh Environment Network (BEN).

Aßheuer, T., Thiele-Eich, I., & Braun, B. (2013). Coping with the impacts of severe flood events in Dhaka's slums – the role of social capital. *Erdkunde*, *67*(1), 21–35.

Azam, M., & Falk, G. (2013). *Governance of climate induced migration in the coastal regions of Bangladesh: New transformation required?* Earth System Governance Tokyo Conference. 28–31 January, 2013, Tokyo, Japan.

Bangladesh Water Development Board. (2011). *Annual Flood Report 2010*. Flood Forecasting and Warning Centre, Dhaka.

Beck, T. (2005). *Learning lessons from disaster recovery: The case of Bangladesh.* Disaster Risk Management Working Paper Series 11. World Bank, Washington, DC.

Beyerl, K. (2010). Der Klimawandel in der psychologischen Forschung. In M. Voss (Ed.), *Der Klimawandel. Sozialwissenschaftliche Perspektiven* (pp. 247–262). Wiesbaden: VS Verlag.

Bonifacio, A. C., Takeuchi, Y., & Shaw, R. (2010). Mainstreaming climate change adaptation and disaster risk reduction through school education: Perspectives and challenges. In R. Shaw (Ed.), Climate change adaptation and disaster risk reduction: Issues and challenges. Community, Environment and Disaster Risk Management 4, 143–169.

Braun, B., & Aßheuer, T. (2011). Floods in megacity environments: Vulnerability and coping strategies of slum dwellers in Dhaka/Bangladesh. *Natural Hazards*, *58*(2), 771–787.

Braun, B., & Shoeb, A. (2008). Naturrisiken und Sozialkatastrophen in Bangladesch - Wirbelsturme und Uberschwemmungen. In C. Felgentreff & T. Glade (Eds.), *Naturrisiken und Sozialkatastrophen* (p. 381–393). Berlin: Spektrum.

Chambers, R. (1996). Participatory rural appraisal and the reversal of power. *Cambridge Anthropology*, *19*, 5–23.

Chatterjee, M. (2010). Slum dwellers response to flooding events in the megacities of India. *Mitigation and Adaptation Strategies for Global Change*, *15*(4), 337–353.

Christensen, J. H., Hewitson, B., Busuioc, A., Chen, A., Gao, X., Held, I., … & Whetton, P. (2007). Regional climate projections. In S. Solomon, D. Qin, M. Manning, Z. Chen, M. Marquis, K. B. Averyt, M. Tignor & H. L. Miller (Eds.), *Climate Change 2007: The physical science basis* (pp. 847–940). Contribution of working group I to the fourth assessment report of the intergovernmental panel on climate change, Cambridge University Press, Cambridge.

Centre for Urban Studies (CUS), National Institute of Population and Training, Measure Evaluation. (2006). Slums of urban Bangladesh: Mapping and census, 2005, Dhaka and Chapel Hill.

Cruz, R. V., Harasawa, H., Lal, M., Wu, S., Anokhin, Y., Punsalmaa, B., … & Ninh, N. H. (2007). Asia. In M. L. Parry, O. F. Canziani, J. P. Palutikof, P. J. van der Linden, & C. E. Hanson (Eds.), *Climate change 2007: Impacts, adaptation and vulnera-bility* (pp. 469–506). Contribution of working group II to the fourth assessment report of the Intergovernmental Panel on Climate Change. Cambridge University Press, Cambridge.

Edris, A., & Collins, A. E. (2010). Cyclone disaster vulnerability and response experiences in coastal Bangladesh. *Disasters*, *34*(4), 931–954.

Gropengießer, H. (2008). Qualitative Inhaltsanalyse in der fachdidaktischen Lehr-Lernforschung. In P. Mayring & M. Gläser-Zikuda (Eds.), *Die Praxis der Qualitativen Inhaltsanalyse* (pp. 172–189). Weinheim: Beltz.

Hofer, T., & Messerli, B. (2006). *Floods in Bangladesh: History, dynamics and rethinking the role of the Himalayas.* Tokyo: United Nations University Press.

Intergovernmental Panel on Climate Change. (2012). *Managing the risks of extreme events and disasters to advance climate change adaptation.* A special report of working groups I and II of the Intergovernmental Panel on Climate Change. Cambridge University Press, Cambridge.

Islam, N. (2005). *Dhaka now – contemporary urban development.* Dhaka: Bangladesh Geographical Society.

Islam, T., & Neelim, A. (2010). *Climate change in Bangladesh: A closer look into temperature and rainfall data.* Dhaka: The University Press Limited.

Mayring, P. (2000). Qualitative content analysis. *Forum: Qualitative Social Research*, *1*(2), article 20. Retrieved from http://qualitative-research.net/fqs/fqs-d/2-00inhalt-d.htm

Mayring, P., Gläser-Zikuda, M., & Ziegelbauer, S. (2005). Auswertung von Videoaufnahmen mit Hilfe der Qualitativen Inhaltsanalyse – ein Beispiel aus der Unterrichtsforschung. In: MedienPadagogik. Retrieved from http://wl3www486.webland.ch/04-1/mayring04-1.pdf, May 27, 2013.

Moser, C., & Stein, A. (2011). Implementing urban participatory climate change adapation appraisals: A methodological guideline. *Environment & Urbanization*, *23*(2), 463–485.

Novak, J. (1994). *Bangladesh – reflections on the water*. Dhaka: University Press Limited.

O'Kane, C. (2008). The development of participatory techniques. Facilitating children's views about decisions which affect them. In P. Christensen & A. James (Eds.), *Research with children: Perspectives and practices* (pp. 125–155). London: Routledge.

Palmer, T. N., & Räisänen, J. (2002). Quantifying the risk of extreme seasonal precipitation events in a changing climate. *Nature*, *31*, 512–514.

Paul, S. K., & Routray, J. K. (2011). Household response to cyclone and induced surge in coastal Bangladesh: Coping strategies and explanatory variables. *Natural Hazards*, *57*, 477–499.

Pretty, J., Guijt, I., Thompson, J., & Scoones, I. (1995). *Participatory learning and action: A trainer's guide*. London: IIED.

Rahman, A. A., & Alam, M. (2005). *Dhaka City state of environment: 2005*. Dhaka: UNEP.

Rajuk Plan. (2009). Dhaka City and Rajuk Plan (Future Dhaka City). 1:20.000. Dhaka: The Mappa Ltd.

Rashid, H. E. (1991). *Geography of Bangladesh*. Dhaka: The University Press.

Reliefweb. (2007). *Bangladesh: Flood migrants pour into Dhaka*. Retrieved from http://reliefweb.int/node/246418. Accessed on Feb 22, 2013.

Rosenstein, B. (2000). Video use for program evaluation, a conceptual perspective. *Studies in Educational Evaluation*, *26*, 373–394.

Sen, A. (1981). *Poverty and famines: An essay on entitlements and deprivation*. New Delhi: Oxford University Press.

Schnettler, B., & Raab, J. (2008). Interpretative visual analysis – Developments, state of the art and pending problems. *Forum: Qualitative Social Research*, *9*(3), article 31.

Schönhuth, M. (1996). PRA (Participatory Rural Appraisal) im Diskurs. *Entwicklungsethnologie*, *5*(2), 11–33.

Schubert, C. (2002). Making interaction and interactivity visible – On the practical and analytical uses of audiovisual recordings in high-tech and high-risk work situations. Technische Universitat Berlin, Berlin.

Schüßler, I. (2004). *Nachhaltiges Lernen – Einblicke in eine Langsschnittuntersuchung unter der Kategorie "Emotionalitat in Lernprozessen"*. Report 1/2004 Literatur- und Forschungsreport Weiterbildung, 27, Bielefeld, 150–163.

Spence, A., Poortinga,W., & Pidgeon, N. (2011). The psychological distance of climate change. In *Society for risk analysis*. Retrieved from http://onlinelibrary.wiley.com/doi/10.1111/j.1539-6924.2011.01695.x/pdf. Accessed on May 22, 2013.

Stalenberg, B., & Vrijling, H. (2009). The battle of Tokyo and Dhaka against floods. *Built Environment*, *35*(4), 471–491.

Thomas, S. (2002). What is participatory learning and action (PLA): an introduction. In *UNHCR*. Retrieved from http://idp-key-resources.org/documents/0000/d04267/000.pdf. Accessed on May 27, 2013.

United Nations. (2012). World Urbanization Prospects: The 2011 Revision, United Nations Population Division, New York.

Walsham, B. (2010). *Assessing the evidence: Environment, climate change and migration on Bangladesh*. Dhaka: IOM.